A Guide to

Competitive
International
Telecommunications

By Gene Retske

BY GENE RETSKE

A GUIDE TO INTERNATIONAL COMPETITIVE TELECOMMUNICATIONS

Copyright © 2002 Gene Retske

Published by CMP Books
An Imprint of CMP Media Inc.
Converging Communications Group
12 West 21 Street
New York, NY 10010

ISBN 1-57820-072-5

Transferred to Digital Printing 2010

For individual orders, and for information on special discounts for quantity orders, please contact:

CMP Books
6600 Silacci Way
Gilroy, CA 95020
Tel: 800-500-6875 or 408-848-3854
Fax: 408-848-5784
Email: cmp@rushorder.com

Distributed to the book trade in the U.S. and Canada by
Publishers Group West
1700 Fourth St., Berkeley, CA 94710

Dedicated to
Lucy Mae Retske,
my mother
May 23, 1914 - January 5, 2002

Table of Contents

Acknowledgments

There are so many people that I owe for their contributions that made this book possible, that it is impossible to remember, or thank them all, so if you feel you were omitted, my apologies. I will get you on the next book. My thanks go to Harvey Starin, Tony Brimo, Glen Benjamin, Rob Bracewell, Jim Hendrickson, Othel Turner, Keith Rhea, Gene Cohen, Chris Hall, Bud Lutes, Glenn Walters, Dianne Costello, and my parents, Myron and Lucy Retske, all of whom contributed to the experiences that made this book possible, either directly or indirectly. I would like to thank Christine Kern for her support and belief, and my wife, Linda, for her contributions, criticisms, praise and encouragement, not only in this book, but in life. Paul Barker, whose layout and illustration work, as well as his friendship, was vital to this book. I would especially like to thank Janice Reynolds, who edited this book, for so many things that it would take too long to mention them all, but her professionalism, motivation and encouragement were greatly appreciated.

Forward

It is difficult to conceive how far telecommunications has progressed in just the past few years. I first entered the world of telecommunications in 1973, when I began working at one of AT&T's subsidiaries, Southern Bell, in Hollywood, Florida. At that time, there was a file cabinet in the file room, which had a bottom drawer that contained the files of *all* the customers who had telecommunications equipment or services *not* supplied by Bell. There were only approximately 50 files in that drawer. (This was out of more than 20,000 business customers that were in the district.) More astonishing, that district was considered to be one of the worst hit by the new wave of competitive telecommunications.

Remarkable as it sounds today, less than 30 years ago that was the state of the competitive telecom market. In 1968, a landmark decision by the Federal Communications Commission in the United States opened the telecommunications market place. In a short period of time this competitive spirit would spread around the world. By 1996, AT&T's nearly 100% market share in long distance had dropped to a less than 50% share.

Today, the winds of telecom competition are blowing in every corner of the globe. Virtually the entire telecom market is competitive and opportunities abound — the technology is exploding. Yet, within this whirlwind there are unprecedented failures, and even the once unconquerable AT&T has reeled close to the brink of disaster.

What are the factors that are driving this shakeout, which some have said is the biggest global upheaval since the fall of the Roman Empire? Surely the telecommunications industry is not going away as a whole!? Will any of the current telecommunications companies be able to survive, or are they all doomed to failure? What will be the factors that will determine the winners? What will these winners provide? How they will make money?

The Guide to Competitive International Telecommunications: Callback, ISR and VoIP is a follow-up to my successful *The International CallBack Book, An*

Insider's View. As an active participant in the telecommunications revolution, I am positioned to bring the events within the competitive telecommunications community up-to-date, and provide a glimpse into the future.

I chose to forego an academic blue sky approach and instead will look at the history and the future of the telecommunications industry from the trenches, where the real battles are fought.

This book covers the history of competition in telecommunications from the Carterphone decision to the impact of the Internet, with an emphasis on learning all the important lessons, so as not to repeat the mistakes of the past. The reader is given a thorough treatment of the technology of telecommunications, from the invention of the telephone to today's global IP networks. Special consideration is given to ensure that the reader understands the implications of the technologies used.

Since much of the history of telecommunications is written in the regulatory and legal frameworks, this book covers the significant actions and issues, so the reader is prepared to see not only how competition has developed, but also where it is likely to lead.

Finally, since the industry has been littered recently with the carcasses of failed startups and crashing incumbents, a working knowledge of how telecommunications companies actually make money is conferred upon the reader. Whatever your interest in telecommunications might be, a comprehension of the financial underpinnings is vital if you are to understand how to assess today's and tomorrow's telecommunications enterprises.

Whether the interest of the reader is in investing, working for, owning or selling to telecommunications companies, this book will provide a complete understanding of the telecommunications industry today.

Gene Retske
November 1, 2001

Introduction

There are a few dates in the history of global telecommunications competition that cannot be forgotten.

1876- The invention of the telephone.

1983- The divestiture of AT&T.

1992- Callback opened markets around the world.

1998- The European Union opened markets around Europe.

2000- The year the stock market bubble burst.

I have been personally and professionally involved in telecommunications since 1973, when I joined "TPC", The Phone Company, Southern Bell Telephone & Telegraph. Working for a public utility like Bell was as close to government service as you could get without wearing a uniform. It was supposed to be a job for life. I remember an orientation class where we spent more time on employee benefits than on the job we were hired to perform.

I was 26 years old and, admittedly, did not have a high level of interest in the detail of retirement programs and service pension calculations. But, I was curious about the technology, and especially, the business of telecommunications.

No one, not John DeButts, the Chairman and CEO of Bell parent company AT&T, nor Don Murphy, my first boss, nor any of the "experts", gurus or oracles that abounded in telephone company offices around the world, saw, or could have seen, what was about to happen.

In retrospect, I cannot begin to grasp the enormity of the events that unfolded in a few short years. Had any of the office pundits been bold enough to predict even a tenth of what actually transpired, they would have certainly been, at a minimum, severely ridiculed. I am not sure what they did in those days with telephone company employees who went completely over the deep end. Maybe they were sent to some obscure Bell Labs location to be the subject of unthinkable experiments, or perhaps they were "processed" into one of the dozens of Western Electric "works", as the plants were called. Now that I think about it, I recall that employees, who were sufficiently mad as to not be able to function in normal society, were frequently promoted to strategic planning jobs at AT&T Headquarters!

But the changes were draconian. Every year brought a new upheaval. Major corporate reorganizations became an annual event, at least. Sometimes, the entire corporation was reshuffled twice in a single calendar year. The phrase, "But they could *never...*" left my vocabulary forever. The unthinkable seemed to become the status quo before you could get over the shivers from the burst of the initial thought.

The breakup of AT&T in 1983 did not really resolve anything. In fact, once the Bell System was fractured, changes and revolutions began to fall like rain from the sky. The various pieces of AT&T, all major corporations in their own right, became each others biggest potential competitors. The competition was fierce, and the market white-hot.

But as significant as the changes were in the US, they were miniscule as compared with what was about to happen around the world. Outside of the US, most countries had combined the telecommunications regulatory authority and the monopolistic national telephone company. The US had always had a separate authority, and a "privately" owned telephone company. Outside the US, working for the telephone company IS government service. When the changes began to sweep around the globe, the earth rocked.

There is an appealing Yin Yang to the telephone business. On one hand, it is a huge business, ripe with opportunity, and niches abound. A minor marketing or technological success can be a major economic success for the entrepreneur. On the other hand, the competitors are large and well entrenched. If you are even partially successful, half the world will jump into your space, and flood you with unwanted competition.

The telecommunications world today is a living embodiment of this Yin Yang. The circuit switched legacy telephone system is far and away the most commonly encountered system, worldwide. This is the Yang, but the Yin of Internet powered, packet switched, elegant and highly functional networks is spreading and proliferating. The economic reality is that both will co-exist for the foreseeable future.

Investment money abounds, and failures proliferate.

The purpose of this book is to help the reader understand what so-called next generation telephony is all about. We are going to see what events got us to where we now find ourselves, the technological revolution that is raging, what has been successful, what has failed, and finally, *how the heck do you make money?*

In the chapters that follow, we will explore the relevant issues and developments that have guided the global telecommunications market to its current state. Based on this foundation, we will project where it is likely to head, and analyze the forces that will take it there. We have an obligatory start with the invention of the telephone, but progress rapidly to trace the roots of the global international telecommunications market openings. To be able to guess what is going to happen, it is necessary to understand what has already happened.

Callback started it. No doubt about it. Look at the extreme reaction that it caused. It was reviled, hated, and viciously attacked by incumbent operators, like the PTTs, US carriers (who provided the underlying service) and regulators. Callback was where the telecommunications rubber met the lip service road. It was a reality. All the lofty ideals and pontification about free telecom markets came slamming against reality when callback started.

You will read about how it works, and why it could be introduced into a closed and tightly controlled market without anybody's concurrence, beyond the callback operator and the end user. You will see how it threatened everyone — incumbent "fat cat" monopolistic national telecommunications carriers, US service providers, telecommunications regulators (including the FCC), UN agencies and even governments around the world.

We will then see how VoIP took the next steps, and has grown to levels that would have been impossible to conceive a few years ago. Today, these "pirates" carry a significant share of the world's telecom traffic, and are growing at unimaginable rates.

Is this a revolution or a mutiny?

Will the world be better off for this, or are we being led down the road to ruin? Is telecom going to re-emerge as a leading technological and financial force, or is it going to fester in a sea of over building for decades to come?

Read on, and decide for yourself!

Section 1

History of Telecom Competition

So some will lead, while others follow.
Some will be warm, others cold
Some will be strong, others weak.
Some will get where they are going
While others fall by the side of the road.

— *From the Tao Te Ching, Lao Tzu, c.600 B.C.*

Chapter 1
Beginnings

When international callback burst onto the scene in the early 1990's, no one, especially me, believed that callback would ever become a serious contender to compete for the very lucrative international telephone traffic handled by the world's incumbent telephone carriers. The decades old system of bilateral agreements, which was supported by the International Telecommunications Union (commonly known as the "ITU"), fostered an artificially high pricing structure for international telephone calls. The national telecommunications companies (mostly government owned) simply set rates for international calls that were often completely independent of the costs involved. This allowed the companies to establish rates for international calls based on what they could get away with — charging consumers for using what was a tightly controlled monopoly service, regardless of the actual costs or value of the service provided.

But, legions of entrepreneurs saw an enormous opportunity, and began to vigorously attack the telecom market. This revolution started with international callback. International callback was especially insidious since it used the existing public network in an innovative way to reduce the cost of making international calls. It was highly innovative, difficult to detect, and almost impossible to stop.

In 1992 I designed a callback system using AT&T products and sold by AT&T salespeople, which only further confused the issue of the legality of callback. Callback, with its gray market technology, offers a considerable reduction in cost to overseas callers, and high revenues and margins to service providers. By 1995, when *The International CallBack Book, An Insider's View* was published, callback had achieved over $250 million in annual revenues. A drop in the bucket percentage wise, but a highly visible demonstration of the virtues of free market competition. That book, and callback, received a lot of attention, and was discussed, and presumably, "cussed" around the world. I became a hero, or a heel, depending on your point of view.

In 1996, Dr. Pekka Tarjanne, then the Secretary General of the International Telecommunications Union, pointed his finger at me following a press conference in Singapore, and said, "I know who you are and what you are doing to us!" At that moment, I knew that callback had succeeded in doing what years of negotiation, legal maneuvering and outright threats had failed to do. Callback had won, and the world's telecom markets were finally opening up to competition.

Callback had slam dunked competition in an arena that was structured from its very beginning to be the province of monopoly operators. Just 20 years earlier in the United States, the monopolistic national carrier, AT&T, had declared that it was a "natural monopoly" and should not be answerable to accusations of anti-trust. Natural Monopoly? And in the United States? Were they kidding? Monopolies could never carry the label of "natural," could they? Didn't it take an incredible series of legal and regulatory supports to maintain this unlegislated position? What is natural about a vast series of artificial supports designed to eliminate any potential competitive threat?

But in the global arena, many monopolistic national carriers actually believed that their positions were beyond "natural." In many of these venues where kings like Louis XIV had ruled, the PTT monopolies acted as if they held their privileged positions because they were imbued by divine inspiration.

The international competitive telecommunications scene has come a long way since its humble, if not tumultuous, beginnings. In fact, there are credible estimates that non-traditional carriers will handle over US$20.5 billion worth of international telephone traffic in 2002! Considering that just a few years ago (1991) callback was barely a blip on the world telecommunications radar screen, this is quite a development.

By anybody's standards, an industry like the competitive international long distance "industry", with over $20 billion per year in revenue is a significant industry. Perhaps it is time for even the most skeptical observers to start considering what this phenomenon is all about.

I hesitate to call this an "industry," because it really is not an industry in itself. An industry is a definable set of products and/or services meeting a specific set of needs and requirements for a group of customers. International competitive long distance does not meet even a liberally interpreted definition of an industry. It is a method for accessing a network where a more traditional method does not exist.

Should callback be discounted because of its humble and sometimes questionable roots?

When Henry Ford started puttering around the Michigan landscape in his motorized carriage, many farmers (who would later reap major benefits from his new invention) were very critical. They saw Ford's contraption as nothing but a reckless nuisance that created noise, caused horses to bolt and endangered the lives of other people in horse drawn carriages, not to mention pedestrians. The horseless carriage was also accompanied by what detractors described as a "very bad odor." (It should be noted that some critics claim that callback also has a similarly repugnant odor.)

Later on, these same farmers found that Mr. Ford's invention could also plow fields at a rate several times faster than horse drawn rigs, and allow access to markets many miles away. They, of course, fell in love with the smelly, noisy contraption.

Likewise, some callback detractors, like PTT's (Post, Telephone and Telegraph) that felt threatened by any form of uncontrolled competition, began to warm to the idea of easy access to networks outside their own borders. One of the most stringent persecutors of callback actually bought an interest in a callback company, and also offered its own callback services in other countries.

Most new industries have similar humble roots. These industries are created by entrepreneurs, inventors and other very creative people with grand, new ideas. But, it is a human trait that the establishment almost immediately rejects new ideas (regardless of how good they may be) for the very reason that they are new.

When MCI first came on the scene in the US in the early 1970's, there were no ticker tape parades. MCI was treated with a great deal of suspicion and skepticism. Today, MCI is not only one of the darlings of Wall Street, but is the recognized pioneer of the competitive long distance industry — starting the competitive telecommunications trend that has continued throughout the world.

In fact, it might be said that the degree of scorn shown callback by the established telephone community — carriers, local operators and government regulatory agencies — demonstrates exactly how exciting and significant this new industry has become.

The concept of a new world telecommunications order is not exactly a radical idea. Those of us in the United States may remember that Bill Clinton and Al Gore made the new world telecommunications order a major issue in their reelection campaign. It is now evident that even seemingly radical restructuring, such as the Telecommunications Act of 1996, doesn't offer enough change, at a fast enough pace, to meet the continuously metamorphose that consumers demand.

The United States is arguably the most liberalized telecommunications country in the world. The domestic and international long distance marketplaces are almost completely open to anyone with the time, money and talent to explore the world's largest telecommunications market. Provisions in the US's 1996 Telecommunications Act have even opened up parts of local service provisioning (previously the exclusive domain of the local exchange carrier or LEC) to competition, which was long thought to be the Lost City of Gold within the telecom community. But, after the initial rush by competitors into the local service market, it has lost its luster, and even well-financed competitors are falling by the wayside. Thus, there is still very little competition in that sector, with LECs controlling 99% of the local service market. So, even in the US, there are many areas of telecommunications that remain virtual monopolies despite all the attempts at liberalization over the years.

With the exception of a handful of countries, the majority of the world has a long way to go before there is a true competitive telecom structure in place. Although there are many sectors in the US where competition has brought benefits, those areas remain untapped elsewhere, i.e. long distance, international long distance, local service, information enhanced services, voice mail and messaging systems and even the equipment used for telecommunications.

Things are changing and changing very rapidly. There are over 200 tele-countries in the world today. Each one has a different set of standards and varying sets of rules and regulations. Within each one of them a different political environment exists that dictates the roll out of true telecommunications competition. The political climate will be the most significant factor in the development of the individual tele-countries' telecommunications competition policies.

Many look at the telecommunications picture in the world today and see a riddle or a problem. Others look at it and see a golden opportunity waiting to be seized. This book explores the position of competitive telecommunications as it is in the world today and what direction it is likely to head — from a rutted dirt road to a superhighway of fast moving ideas, any one of which the business community may appropriate as the next great opportunity.

Chapter 2
126 Years of
Telecom History

It is a standing joke among former Bell System employees that every Bell System business presentation began with "In 1876, Alexander Graham Bell Invented the Telephone." Despite the risk of being accused of perpetuating that tradition, I think it is necessary to summarize the relevant parts of telecom's history, in order to set the stage for understanding the global revolution that is taking place today.

In 1876, Alexander Graham Bell invented the telephone. (I got it in, now I am fulfilled!) But as significant as this invention was, it has to take second place to the invention in 1901 by Theodore Vail, who invented the telephone *industry*. Behind every great inventor is an entrepreneur, and Vail, who is far less well-known than Bell, is more responsible for the spread, structure and evolution of the telecommunications industry than Bell himself. Vail had a secret formula and he used it to establish the Bell System's position.

At the time that Bell invented the telephone instrument, there were several other inventors hot on the heels of this invention. In fact, in 1976, Bell almost lost the patent on the telephone to Elisha Gray, who filed a caveat indicating that a patent filing was forthcoming on the exact same day, February 14, 1876, that Bell filed his patent. Although Gray's filing was only a few hours after Bell's, Bell won the dispute on the basis of the slight time differential, and the fact that Bell had described a working device and Gray had not.

Vail quickly recognized the potential of Bell's invention, but also knew that it was highly vulnerable to competition. To secure a favored place for Bell (the

Figure 1. New York Telephone Spaghetti Bowl.
Property of AT&T Archives. Reprinted with permission of AT&T

man and the company), Vail moved to define the service and the device itself as a utility, and suggested that local authorities create regulatory bodies to oversee the operations of the telephone company, thus "protecting" the public. What Vail was really trying to protect, however, was the monopolistic structure of this new industry, by locking out Bell competitors. This was accomplished through the use of his secret formula.

What was this "secret formula" that Veil concocted?

To go into a community, suggest that the government closely regulate telecommunications, and offer local governments a new service to tax!

In addition to adding to the local government's coffers through taxation, Vail also offered free telephone to the local governments in exchange for an exclusive "franchise" to offer telephone service within that community. Not a big premium by today's standard, but back then it was a valuable commodity. This was in addition to receiving a new revenue stream in the form a taxable service. In the state of Florida in 2001, these taxes can be over 20% of its total tax revenue.

One of Vail's most powerful tools was a picture, taken in New York City, which depicted a spaghetti bowl of telephone lines, so many that they nearly obscured the sky. At the beginning of the 20th century, the US was still, by and large, a rural agrarian society, and the notion of a plethora of metal wires overhead, and ugly, naked telephone poles looming large over one's head was abhorrent.

Thus, Vail, and Bell, received an exclusive, nearly perpetual license and a guarantee of no competition — a veritable lock on the telecom market for the critical first years of the technological revolution, and a guaranteed fast start for the Bell System. Outside of the US, the same model of a monopoly was used, with a slight twist. The national governments owned and operated the monopoly company and ran the regulatory authority as a single, self-regulating entity, often carrying the British moniker of Post, Telephone and Telegraph. In the US, AT&T's supposed undue influence over its regulators was likened to "a fox in the chicken coop." In most other countries, however, the fox literally owned and operated the chicken coop.

For well over a century, the concept of the telephone company as a "natural monopoly" was embodied in the slogan for the Bell System, "One Company, One Policy, Universal Service." If you look at the old Bell System logo, you will see this inscription surrounding the Liberty Bell Symbol.

This secret formula led to the Bell System's eventual dominance within the US and eventually, globally, as Vail's secret formula became copied around the world. It's the model that still underlies the global telecom structure.

Vail's model of voluntary regulation, in exchange for a monopoly service, formed the basis for the establishment of telephone service throughout the rest of the world. It remained in place, virtually untouched, until 1968, when the FCC made a critical decision to allow the connection of non-Bell provided equipment to the Bell network. This change, which at the time appeared to be tiny and insignificant, ultimately laid the legal and regulatory basis for the liberalization of ALL telephone monopolies.

From 1968 to 1983, a steady series of US legal decisions (by the courts and the FCC) began the inevitable process of opening up more and more of the fast growing, lucrative telecom market to a burgeoning group of competitors. Each eager to gobble up the crumbs falling off the table of the world's largest company, which was now called AT&T (another great Vail inspiration). This process was watched with a jaundice eye, if not downright hostility, by the PTTs around the world, who viewed this as a second coming of the American Revolution that had spread and toppled the great monarchies of the 19th Century. Their fear was that the PTT cash cow might meet the same, often ignoble end, which befell many European monarchs.

The PTTs were right, 1983 was the telecommunications equivalent to 1776 in world politics. But for many reasons, it was also the bellwether year in telecommunications. For instance, the world's largest company, AT&T, was split up into 9 different pieces — AT&T, the 7 Baby Bells, and Bellcore (the former General Departments of AT&T, now Telcordia). Each representing a major corporation in its own right. But also, "equal access" allowed others to freely compete for long distance services, which presented AT&T with a structural challenge to its dominance of the world's largest telecommunications market.

These were the first steps in a process of divestiture, segmentation and even decline; not only for AT&T, but also for its equivalent companies throughout the world. AT&T was, however, an exception among the brotherhood of PTTs, since it was the only one that:

* was a completely privately-held company,
* lived a separate existence from the national telecommunications regulatory authority,
* had to publicly report its annual operating revenues and expenses, and
* had any need to answer to the public at large.

It was also the first one to lose its exclusive franchise to provide telecommunications services.

For all these reasons, the US experience was closely watched by the rest of the world. Not just because of the novel nature of the goings on, but because the telecom industry was aware that what happened here would form the basis for a worldwide telecommunications revolution, like it or not. There would be many variations on the theme, but like a tidal wave that cannot be stopped or slowed until it runs out of water to swell, the telecommunications revolution will not stop until the last citizen on the face of the earth is affected.

Chapter 3
The Quiet
Revolution

The telecommunications revolution that's taking place today rivals the Industrial Revolution in its potential impact on mankind. Much has been written about the Information Age, and the incredible explosion of information systems to manage and move this information and the business opportunities afforded by this development. For many people, the word "Internet" in a public stock offering, or a newspaper headline, is an "attention grabber."

One of the leaders of this revolution, however, had a quieter, almost stealthy, beginning, although it did emerged to spread the revolution around the world. Its name is "Callback." However, even the author, who has sometimes been called "The Godfather of Callback," has to admit that even he didn't give *that much* importance to callback. Callback is not an industry, not a market, and not even a business. In reality, it is an access method that is almost silly.

All the same, this "silly" technology is what fueled the fires of the telecommunications revolution. Even Dr. Pekke Tarjanne, the Secretary General of the International Telecommunications Union (ITU) has credited callback with causing the downfall of the accounting rate system. It's this antiquated system, as we shall see later, that promulgated the telecommunications monopoly for the better part of the 20th century. This monopoly was, and still is to some extent, responsible for stifling the spread of affordable telecom services to the citizens of the world, promulgating a "have/have not" scenario, where only the higher economic strata can easily access telecom services. The economically privileged enjoy the benefits of the technology, and the unwashed masses become further alienated. Let me assure you that I am far from being a Marxist, but the withholding of the advantages of the information society from lower economic strata is an undeniable result of closely controlled telecom resources.

The artificially inflated price structure for telecom services, particularly international long distance, created a pent-up demand for any kind of affordable service that attracted the interest of entrepreneurs, who saw the potential for profit. At the same time, the existing technical and financial structure of the global telecom industry offered the possibility for exporting the relatively low US international long distance rates to users around the world.

It is impossible to deny that callback, which transverses unauthorized and unapproved through a "guaranteed" telecommunications monopoly, upsets the order of things. Once the competition starts, a series of events, leading to the destruction of a system of artificially inflated international calling rates, is irrevocably set into motion.

☎ Chapter 4

The State of the
New Telecom World

The first order of business is to bring everyone up to date on the state of telecommunications structure in the world today. This isn't easy — there are many different countries involved — over 250 "tele-jurisdictions" currently exist, each in a different state of roll out, at different rates of speed, with different perceived final results. The European Union (EU) and the World Trade Organization (WTO) have tried to impose some degree of standardization in this chaotic scene, but to little avail.

There are generally four categories of liberalization strategies in the world today:

* Countries that are liberalized (a very small handful).
* Countries that have liberalized some of their telecommunications structure and are moving toward liberalizing the rest of it.
* Countries that have announced that they are going to liberalize part of their telecommunications structure but have done nothing about it.
* Countries that have announced or decided that they are not going to liberalize any of their telecommunications structure.

It is important to note that regardless of what we want to think, not every country perceives the same end result as the goal of competition. It has been stated in major publications that the total liberalization of the telecom marketplaces is a foregone conclusion and it is just a matter of time before all countries will end up with a liberalized environment. This is a very naïve and incorrect point of view. In fact, the countries that intend no liberalization and intend to leave things "as is" are in the vast majority.

Although believers in free market economics have no doubt that the more reluctant countries will ultimately be pushed into coming in line with the structures of free market telecommunications, there are many others who believe that regulatory and legal barriers will prevent this from ever occurring.

What is certain is that there is a great deal of confusion, not only among telecommunications' minor players but among the major players as well. It is also clear that something is happening; the status quo is changing — technology is pushing it, the availability of network resources is pushing it, and also political and economic trends around the world are pushing it. All of these factors are stimulating changes within the telecommunications industry, although it's not clear exactly what the final structure will look like, if indeed, that is any significant change at all.

The ITU, the major telecommunications coordinating body for the world, has for several years carefully avoided taking a stand on this issue. Many of the trends, policies and changes in the ITU's regulations over the last several years show that very little change is actually being contemplated; although it's noted that there hasn't been any clear course of action or general policy direction decided by the membership at large.   Thus, it's impossible to determine whether this is a deliberate attempt to hold onto the existing structure as long as possible or if a deliberate decision has been made to try to stifle progress, or even whether this inaction is the result of a political impasse within the ITU. What is certain is that decisions this monumental, far-reaching and politically controversial will be taken very slowly, if at all.

The bottom line appears to be that anyone who is waiting for the ITU to take a position of leadership or even take a stand one way or the other on the issue of liberalization is probably waiting on the improbable. It is much more likely that the ITU will continue to do nothing as long as it is not forced to take a stand or make a decision. More likely, the future course of telecom will be decided by external technology and economics. Some companies may fall by the wayside, be acquired or simply go out of business, if they cannot effectively compete and build markets rapidly enough.

Chapter 5
The Great
American Experiment

Much, probably too much, has already been said and written about the divestiture of AT&T, the breakup of the Bell System and their influence on US long distance competition. Most of us will choke if we have to listen to the recounting of the miracle of MCI one more time, although we are still arguing about the long-term impact, some of the details and who really benefited.

Telecom Competition History 1968-2001	
1968	Carter Fone Decision
1969	MCI Specialized Routes
1970	First Nationwide Alternative Carriers
1971	SCC Decision / MCI Commercial
1972	Southern Pacific Network
1977	Execunet Decision — Switched LD
1979	Computer Inquiry II
1981	Resale Decision — Equal Access
1982	FCC Deregulates OCC's
1984	Equal Access Allocation
1985-1989	Quality-Value-Service Competition
1990-1995	Commodity Competition
1996-1999	Global Competition Begins
2000-2001	Shakeout Starts

Figure 2. US Resale Competition Major Events.

What virtually no one is arguing is that the US has now achieved a high availability of communications services at rates that are among the cheapest in the world. The carriers are not hurting either, just look at where AT&T stock had risen to before the big telecom stock crash. AT&T went from 17+ in 1984 to over 60 in the late 1990s. Even with the depression in the stock in the 2000s, it was still in the neighborhood of 60, considering the various spinoffs — Lucent, Wireless, etc. There are those who still try to argue that all this could have happened anyway, at less cost and less pain. There are also those that argue that the South really won the Civil War!

In order to understand where the telecom competitive environment is headed, it is worthwhile to one more time rehash the restructuring that has occurred within the US telecom market since 1968 and to relate it to the trends that callback started throughout the world. Remember: Those who fail to learn the lessons of history are doomed to repeat them.

The US trend began with a Supreme Court decision in 1968, which stated that the telephone company could not deny connection of a device based solely on its use. Prior to 1968, AT&T had always maintained that it was the legitimate guardian of the public telephone network. By this decision, the Supreme Court opened the possibility that AT&T may not be the only possible overseer of the network. The network began to become truly public.

This led to an environment where MCI could (and did) provide private line services between select routes and ultimately connected those calls to the public network. It was argued, successfully, that the "public interest, convenience and necessity" might be better served if there were other, specialized carriers.

In 1970, Datran asked the FCC for the right to set up a nationwide network for data only, the FCC invited domestic satellite proposals, and MCI asked permission to link its three microwave systems.

By 1971, the FCC's Specialized Common Carrier decision officially opened up private line competition. In less than three years, the US telecommunications industry had gone from a monopoly to a free market model — at least in private line services — and this was only the beginning.

In 1972, the FCC "Open Skies" ruling created an entire industry — domestic satellite communications carriers. And it gave Southern Pacific (soon to be Sprint) approval for an eleven state common carrier network, although only for private line services.

To offset real and perceived market erosion, AT&T filed a "high-low" tariff in 1974. AT&T asserted that MCI, Southern Pacific, IT&T and others were "cream skimming," i.e. providing service in only the areas where traffic, and thus profits, were high, while AT&T was obligated to provide "universal service." The high-low tariff was designed to provide low rates in high traffic routes and higher rates in lower traffic routes. This was AT&T's first competitive response and it set the trend for the future.

Competition in the specialized common carrier arena, i.e. private line service, rolled right along and AT&T did not go out of business. Then, in 1977, the first watershed event occurred — the FCC issued its Execunet decision. This decision allowed Execunet, MCI and the other specialized common carriers to provide switched long distance service in addition to "specialized" carrier services.

There is a direct analogy to callback (and VoIP) here. It's said that in the beginning of domestic resale in the US, the carriers had to "sneak in the back door," since the front door remained firmly shut. Callback and VoIP have been forced to sneak in the back door internationally, for the same reasons.

In the 70s, the real issue in the US was long distance competition, AT&T knew it, the FCC knew it, and the public should have known it. Providing private line services at a discount rate, especially since a lot of the facilities were leased from Bell, was not easy or highly profitable. I worked for an AT&T subsidiary, Southern Bell, at the time. We were told not to worry, MCI would go broke competing for the private line market, and the FCC, the Congress and the Courts would never allow a competing, parallel network to be established.

Today, we are told that callback is a transitional technology, and will go away when all telephone companies in the world are privatized and the existing long distance carriers take over. A Wall Street banker once said that if all the countries in the world agreed to a single rate structure, callback would be out of business.

If all the countries in the world got together and agreed to get rid of their armies, there would be no more war, either. And there is probably a greater chance of getting nations to abandon armies than there is of them agreeing to a common rate structure. Callback is here to stay.

Anyway, back to historical perspective. In 1978, the Justice Department filed an anti-trust suit against AT&T to force the divestiture of the company. In the same year, MCI received approval to expand Execunet, which it had acquired.

The AT&T divestiture case garnered most of the attention for the next few years as its competitors grew, strengthened and consolidated. AT&T was functionally broken up in 1983, and formally split into nine pieces in 1984 —

AT&T itself, seven Regional Bell Operating Companies (RBOC) and Bellcore, a company that coordinated the network and set network standard for the RBOC's.

Prior to 1984 and the institution of equal access, using an Other Common Carrier (OCC), such as MCI, was a royal pain in the neck. To make a long distance telephone call, you had to dial the provider's 800 number, input an 11-digit account code, then a 5-digit PIN, before dialing the telephone number that you wanted to call. This was a total of thirty-nine digits, as compared to the eleven that you dialed for an AT&T handled call. It was not terribly convenient, or reliable. The 800 numbers were often busy (sometimes for hours), blocking any attempt at placing a call. And the billing was, well, atrocious. But, despite everything, the users decided that all the trouble and inconvenience was worth the money saved, and MCI and others grew and thrived.

Today's callback is not much more convenient, and has some of the same problems, and VoIP adds a few new ones. Both certainly have their share of billing problems. But just like in the early days of US resale, users have decided that the problems are worth the effort. The savings are comparable to those early US long distance days, and for the provider it's often even more profitable. And, as with the US experience, the service is growing by leaps and bounds.

The world economy is in a similar position to the US economy in the early 1970's, with expansion of markets and a proliferation of competitive threats. But, instead of building *nationwide* marketing and distribution channels, we are now building *worldwide* channels.

In 1984, telephone companies were required to provide one-plus, equal access to carriers. Our history will end at this point, because the rest is the saga of free market marketing.

There is a danger in drawing too close of an analogy between the experience in the US and what is occurring in the rest of the world today. There are obvious differences between how one authority, the FCC, makes a decision for a competitive market in the US and how two hundred plus PTTs will face the same decision in their market.

Despite the similarities, there are cultural differences that run wide and deep. And in many regards, the global economics in the 1990's and the US economy in the 1970's are very different. In fact, there are probably more differences between the US telecommunications environment of 1970 and the global environment of the 1990's than there are similarities. So, it is probably completely illogical to compare the two...but it is very tempting to do so. Even irresistible. To ignore the similarities because of the differences would cause us to miss some of the most important points about resale in general, regardless of when and where it occurs.

One major point to remember is that the goal of the US resale struggle then and VoIP today are very similar in many regards, because they have the same objective — full and open competition for the long distance traffic of a nation. The history of international resale, which will be written in the next few years, will show that callback was the driving force in the battle for free market competition.

The telecommunications market openings that callback started have now been accelerated by VoIP. Now that a cost-based competitive model is possible, we should see prices begin to stabilize, and even more markets thrown open to competition.

Chapter 6
International
Callback

Why did callback have such an impact on the world telecommunications order? It barely reached a measurable market share, and was used by such a small segment of the world's population, that it was virtually undetectable.

Like many early trends, its importance is not measured in market share, but in mind share. The fact of its existence was far more significant than the revenues it impacted. In fact, a very good argument arose out of the Argentine callback wars, where the existence of callback actually increased PTT revenues and stimulated demand for long distance services.

The argument says that there is a certain, very large percentage of international calling that is going to occur, regardless of the price point. It simply HAS to go. In a traditional PTT model, this "must do" traffic accounts for the vast majority of the total international traffic. Next is the second tier, which consists of discretionary traffic. If the cost of the call is too much, then letters (and presumably, email) will HAVE to serve the purpose; however, if the price is attractive, some of the traffic may switch to long distance. Finally, there is a third tier of traffic that will go long distance if, AND ONLY IF, the price is low enough to be perceived as a great bargain.

What callback did, for the most part, was to stimulate demand in the second and even third tier traffic.

Immediately after callback opened up telecom markets worldwide, entrepreneurs began to discover the

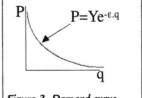

$$P = Ye^{-\varepsilon.q}$$

Figure 3. Demand curve.

realties. What happened next was not a pretty sight, but is a textbook exercise in commodity-based competitive markets. It is only natural that in a free market, competitive pressures will take over and begin controlling the pricing and availability of products in the marketplace.

Let's look at an example. In 1995 the Brazilian PTT's (Embratel) rate for calls to the US was $2.25, with a 3-minute minimum. At the same time, because of competition in the US, carriers were selling wholesale minutes from the US to Brazil for less than a dollar (typically around 75 cents).

This was a huge arbitrage opportunity. For example, if a caller in Brazil used the primary competition, Embratel, to send a one page fax to the US, that caller would pay $2.25 AND a 3- minute minimum, for a total cost of $6.75. Now compare that to using a callback operator, who could send the same fax for 90 cents (75 cents for the US to Brazil leg, plus 15 cents for the US termination).

The first callback operators into Brazil began offering callback with rates ranging from $1.90 to $1.60 per minute, usually with a one-minute minimum. Even at those rates, the discount to the consumer was enormous. A one-page fax could be sent at over a 60% discount of what Embratel charged. Thus it began, and the world of unregulated competition was launched.

Even though a minimal number of minutes, like 10,000 per month, could bring in over $20,000 with margins of 50-75%, callback operators began equipping their switches to handle greater quantities of traffic. A single T-1 to a US carrier could handle in excess of 300,000 minutes per month, and most equipment was sold in T-1 increments, so why build a business for 10,000 minutes when you can just as easily accommodate 100,000 or even a million minutes?

Callback operators — dreaming of immense volumes of traffic, all at high margins, and highly profitable — geared up for what appeared to be a major opportunity. They built their networks to match their dreams: a massive volume of calls. In an effort to get better pricing, they also committed to an ever-increasing level of international calls with their US carriers. But, they failed to take into account the very nature of callback — it's a gray market service, at best, meaning that traditional mass marketing and distribution techniques were not generally available.

Early practitioners could not use television, radio or print media to advertise their gray market services since these services were not officially sanctioned by the state owned and operated telecom company. Additionally, the base of customers was limited to those who could, or would, take the risks that might be involved with their use of callback (regardless of how attractive the savings might first appear to be).

Given the stealthy nature of the business, the most prevalent sales method involved the use of independent sales agents to sell the service to customers. Typically, these agents shared richly in the sale of such service, to the tune of 15% or more of the gross receipts as their sales commission. Entrepreneurial companies sprung up based on this opportunity alone. Since mass marketing techniques were not generally an option, the universe of callback users was somewhat limited in size. After the initial sales effort, which usually involved several companies, this limited market became known to most of the practitioners in a geographic area.

Due to the lack of sophistication in the sales and marketing effort, the primary differentiating factor, in fact the ONLY differentiating factor, promoted was the per minute price for calls. Thus sales were made based almost entirely on price, and the customer, knowing no better, determined the value of the service based on that factor alone. Through these unsophisticated marketing and sales efforts, customers were actually conditioned to make their choice of callback companies based only on the price. This meant that, if the customer were offered a lower price for an apparently equal service, they would take it.

Callback operators, who had made significant investment in network, equipment and business operations, quickly realized that they could easily capture a bigger market share by reducing the unit price. Starting in 1995, companies like Kallback, Viatel, IDT, USA GlobalLink, USFI, and MTS, who had substantial financial backing and/or resources, began to aggressively price their products, to the detriment of smaller operators who were less well financed.

What happened next was not a very pretty scene. Over the next two years, the unit price for callback services in countries with the biggest arbitrage opportunities began to plummet, to the delight of callback customers. Operators that had once charged $1.80 for Brazil to US calls were dropping the charges down to $1.20, then to under a dollar by 1997.

Since callback relied on the use of existing PSTN services offered by US-based long distance carriers, the callback operators began putting additional pressure on those carriers to lower their "wholesale" rates. Now, these rates were typically not based on any actual cost model for providing the service, but on a series of rather arbitrary bilateral agreements between countries. There was considerable room for maneuvering within the bulk rates the carriers charged for services, if the carrier was creative.

For example, it was possible to actually sell a minute to Brazil for less than the accounting rate to Brazil, and still make money. In fact, lots of it. See the chapter on Accounting Rates for a complete discussion of how accounting rates work.

How did all this work in allowing callback to emerge to the point that the head of the ITU recognized what had happened and pointed an accusatory finger at the author?

It is simple. Callback actually was the death knell of the bilateral, take-what-you-can-get, agreement system fomented and sustained by the ITU and its member nations. Mainly because that system is based upon a financial model that has a fundamental flaw — it's based neither on the actual cost, nor on the consumer value of the service, and can only exist in a very tightly controlled market where no competitive product is allowed.

Once a cost-based alternative to monopoly service appeared (callback), the artificially inflated arbitrary pricing of the ITU's accounting rate system fell apart. When a consumer is offered the opportunity to choose an alternative product, the monopoly product is faced with a potentially devastating situation.

If a monopoly PTT could not control the pricing available in the market, and consumers could select an alternative service, the PTT would lose revenue rapidly — at least in theory. Case studies suggest that overall traffic levels actually increased when callback came into the market. But, for varying reasons, this did not necessarily help the PTT.

First, the bulk of the margin in international long distance calls is typically in the collection rate, not in the accounting rate. When international traffic in a particular country shifts from outgoing traffic, where the PTT receives the collection rate, to incoming traffic, where the PTT receives only the settlement rate (half the accounting rate) from the other end, it loses margin, if not revenue. In most cases, revenue will increase as callback traffic increases, but the margins will decrease dramatically.

PTTs recognized this and began a legal, regulatory, technical and marketing attack on the unwanted competition, wherever it appeared. This "spy vs. spy" battle often led to humorous consequences.

In one instance, a callback company in the US Midwest had a high volume of traffic to a certain South American country. This country decided to prevent the callback service provider from receiving code calling attempts. So, they blocked all calls from that entire area code — it was one that was not called often.

As a countermeasure, the callback service provider got a group of DID numbers that were not only in Washington, DC, but in the exact same exchange as the country's embassy. Blocking those calls would have shut their embassy off completely from any calls from the government of that country!

A further consequence of callback traffic increase was the infamous unanswered, and thus, unpaid for, trigger call. The PTTs, during the FCC hearings on the callback issue in 1995 and later, contended that the economic impact of the "avalanche" of incoming trigger calls was enough to flood the network and put most small PTTs out of business. On the other side of the issue, the US-based callback companies said that the calls were of short duration, and had a minimal impact. Besides, they reasoned, not charging for unanswered calls was more of a marketing decision anyway, and that since no PTT tried to charge for them, they were throwaways.

The reality is probably somewhere in between. It's difficult to imagine that trigger calls, which are very short in duration and do not require a full call setup, could have a serious impact on a national network.

How people came down on this issue was pretty much dependent on how they felt about competition in general.

The battle against callback provided a mixed bag of results. There is little doubt that callback grew, and grew rapidly, to the tune of sometimes as much as 100% each year. But, the constant barrage of legal, regulatory and technical attacks took its toll, and callback has yet to reach its full potential, thus attesting to the limited success the PTTs had in suppressing its usage. In all, you could probably score the round to the PTTs, which mostly only wanted to delay the onslaught of inevitable competition.

Yet, the genie had been let out of the bottle, and users around the world began to see and experience the oft-touted benefits of a free economy and the ensuing competition in telecommunications. The taste was often sweet — lower prices, better services, more features — and the consumer appetite was whetted for more. And technology was poised to spring a new round of intrusions into the private domain of the PTT.

Chapter 7
The Truth About International Simple Resale

International Simple Resale (ISR), a logical extension of callback, was the next wave in international telecommunications competition. Callback, you will recall, involved the use of already created minutes and was a pure arbitrage play, relying on the difference in price for identical call paths. Let's recap since this is an important issue. Callback exploited an anomaly caused when two national carriers, sharing the cost of an international transport facility, set their *collection* rates independent of each other.

For example, if Country A and Country B construct a telephone circuit between them, they will hypothetically share the cost to build and operate it. They will then agree on a formula for one country to compensate the other in the likely event that the amount of traffic that Country A sends to Country B and vice versa is not exactly equal. (In fact, it almost never is!) This is known as the *accounting rate*, and it establishes a *settlement rate*, which is equal to half the accounting rate. The settlement rate is the price per minute that the country sending the most traffic pays the country with the lesser volume.

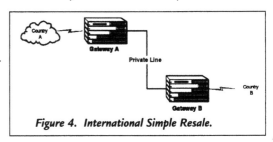

Figure 4. International Simple Resale.

In a perfect world, the accounting, or settlement rate would determine the price that each national carrier charges its residents for the calls (the collection rate), but it does not. In fact, historically, the collection rate is quite independent of the settlement rate, which may only serve to set a minimum for the collection rate. It goes without saying that the accounting rate was more often than not set without regard to the *cost* of providing the service. In the past, the accounting rate was *far* in excess of the cost. The excess charge represented an enormous margin to the national carrier, who was free to charge whatever they wanted for the service.

When callback entered the scene, it used the difference in the rates charged to establish an arbitrage system. Think of the irony involved. If Argentina charged $5 per minute for a call to Miami, as we cited in *The International CallBack Book*, and the US carrier charged only $2.50 for the *identical network facilities*, there was a tremendous opportunity created. This is where callback began its rise, which, in turn, sharpened the public's appetite for reasonably priced telecommunications services. All callback did was to reverse the direction of the call. Nothing more.

But as competition heated up between callback carriers, and their traffic levels grew, those same callback carriers began to focus on the actual *cost* of the facilities that they were using. It took very little effort to realize that they could create the minutes for far less than they were paying for them, and the push for ISR had begun.

ISR offered very competitive rates, well below the best price offered by the lowest priced licensed carrier, and at a quality of service that was often better. ISR was clearly an important new development in competitive telecommunications.

There was only one, very minor, problem with ISR. It was illegal in nearly every jurisdiction!

Not that this technicality stopped the most determined tele-competitors. In fact, this type of barrier to competition made it even more attractive. The easy availability of satellite connections to nearly every tele-jurisdiction made it possible to sneak into a country virtually unnoticed. Especially if the connection was of a non-traditional nature, like VoIP. With the right kind of connections, in-country traffic could run both ways. As you will discover later, traffic *originating* from a tightly controlled tele-jurisdiction is much more profitable than traffic *terminating* to that country.

This whole scenario was complicated by one of those ironies that always seem to crop up when you least want it. On one hand, if the business was successful, the government regulators might well discover it and shut it down, or worse. On the other hand, if it failed to generate a sufficient level of traffic to have much of an impact on the national carrier, it could go unnoticed for a long time. It made for an interesting balancing act!

Chapter 8
Least Cost
Routing Carriers

OK, I admit that this is my own classification, but it has relevance, so I am going to use my license as an author, and define this space. It is really a modification of what used to be called a "reseller". The basic concept is to set up a switch, billing software and a provisioning system that will allow the reseller to connect to carriers with networks and to resell their destinations. Callback companies are essentially resellers, with a specialized access method. They connect to one or more facilities-based carriers and offer their destinations, obviously marking up the cost for the routes in the process.

Today, there are legions of resellers, arbitrage carriers, and boutique operators who interact with, supply to, and even operate as "least cost routing carriers". It is this model that has caused the biggest impact in opening up the telecom markets around the world. They specialize in providing less expensive communications where the consumer demand exists. It is the purest form of entrepreneurialism.

A so-called "boutique operator" is a small carrier (most next generation carriers tend to range from small to very small) that specializes in a direct route to a specific destination. The classic example is a guy at 60 Hudson Street in New York, NY who has a business partner in a foreign country. (60 Hudson Street is the address of one of the US's most famous carrier hotels, i.e. an independently owned and operated facility where carriers, large service consumers (e.g., ASPs), and resellers co-locate their POPs so they can use premises-based

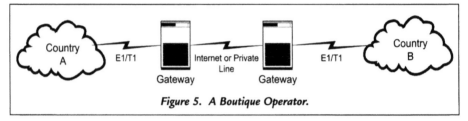

Figure 5. A Boutique Operator.

cabling to interconnect with one another.) They find a technological way to overcome all the practical obstacles to creating a direct line between the two countries. The legalities are circumvented, or ignored, and traffic begins to flow, usually predominately in one direction or the other.

Many other types of carriers also fit this profile, including prepaid calling card companies, and resale carriers. In fact, to make a truly confusing concept almost inconceivable, it turns out that today nearly all carriers are also resellers, at least to some extent. Even the big guys AT&T, MCI WorldCom, Sprint, BT, DT, etc. buy capacity from other carriers, and from each other.

The primary characteristics of a Least Cost Routing Carrier (remember this is my definition!) are:

Physical Plant: They will have at least a switch, or a partition, which they use to make the connection between the end user and the network facility. This switch is usually based on circuit switched technology, but may include the capability to connect to packet or IP-based providers.

Branded Product: The Least Cost Routing Carrier brands the product offering as his own, even if he has no facilities of his own.

Retail Orientation: This is admittedly the loosest of these characteristics, and the lack of a "retail" customer base does not necessarily exclude the carrier from this qualification. But, in general, since the Least Cost Routing Carrier is reselling someone else's network, it becomes increasingly difficult to compete as a middleman in a highly competitive marketplace.

Use of Legacy Networks: Meaning that the carrier usually provides circuit switched services. It is difficult, although not impossible, to compete with the lower cost structure and easier connectivity of the IP-based carriers/clearinghouses.

The theory of a least cost routing carrier is that you will get the lowest overall rates, because the carrier is able to find and provide the lowest rate in the market for any particular destination, and can, therefore, give you the best rates around. Unfortunately, it doesn't really work that way, and most heavy users, whether they are resellers, long distance companies or large enterprise customers, find that they must connect to more than one carrier to get the best possible overall rates.

There is a school of thought that says, the more you buy from one supplier, the bigger the discount. This was a marvelous idea in the 1980s and into the 1990s, but the economics of the situation never really proved the theory to work in practice. It was more of a marketing gimmick than a financial reality. Why should anyone think that there is any reason that buying over a certain number of minutes to Brazil would qualify one for a discount on calls to China? There is really no reason, except to try and maintain as much of the customers' business as possible.

The advent of telecom commodity exchanges will undoubtedly spell the end of the bulk discounts forever.

Chapter 9
Refile

There has been a great deal of talk and discussion about refile, and a few companies have even announced "their refile" as if they invented it. At the ITU gathering in Singapore in June 1997, refile was discussed — not only on the floor of the exhibit halls, but in the sessions themselves. Dr. Pekka Tarjanne, Secretary General of the ITU, spoke about it in his opening press conference, and the theme repeated itself throughout the ten days of the show.

USA Global Link held a press conference at that same ITU gathering to announce its "global refile" service. I am not certain what the point was, since they would not disclose any details — volumes, countries with whom they have agreements, or any details of the technical arrangement — the press conference appeared to be a fishing expedition. Refile is certainly not something that they invented, or something they were the first to use. Nor is it likely that they are a major player in the game.

What is Refile?

The intent of refile is to reduce the cost of carrying traffic from one country to another by taking it from the origination country to the destination country by way of a third country. The reason for doing this is usually to reduce the cost of the traffic because the cost from the origination country to the "Trojan horse" country, plus the cost from the Trojan horse to the destination country is less than a direct route. Of course, in a cost-based accounting rate system, this would likely not succeed, since the actual costs of this arrangement would often be more than a direct route.

Refile isn't to be confused with transit agreements, which are approved methods of carrying traffic from one country to another where a direct route does not exist. Transit agreements are very legitimate and necessary arrangements, and are covered by ITU regulations. In such a case, the intent of transiting the traffic through the neutral country is not to arbitrage arbitrary accounting rates, but to make a "virtual" route where an actual one does not exist.

The key difference between refile and transiting is the intent.

Transiting is the routing of traffic through a third country with the knowledge and approval of all countries involved. There are myriad reasons this is done. The most obvious is where facilities for the direct routing of traffic from the origination country to the destination do not exist, or are inadequate. These are very common agreements, traditionally made between the national carriers in each country. The traffic carried is pre-approved by all countries involved, properly registered ("filed"), and subject to the same terms and accounting rates as the existing agreement between the origination and destination countries.

Refile, on the other hand, is the deliberate rerouting of traffic through a third party country for the purpose of bypassing accounting rates specified in the operating agreement between the origination and destination countries. Normally, the reason to refile is to avoid high accounting rates in the destination country, meaning that the destination country does not approve of the arrangement, and does not benefit from it.

Refile seeks to exploit loopholes in the agreements between the origination and the destination countries, and the third party and the destination countries. The key loophole being the difference in accounting rates.

Like callback, if accounting rates were based on the actual costs, the arbitrage opportunities would be very limited. As it is, artificially inflated accounting rates create a smorgasbord on which traffic brokers and resellers can graze. Even more perplexing is establishing whom the aggressors, beneficiaries and victims really are. One of the most common uses of refile is the origination country refiling some of its own traffic to get a higher margin to a specific destination country, thus violating its own operating agreement.

While there are many variations on the refile theme, there are two basic ways in which traffic from one country can be routed through a third country, i.e. refile, outside of a transit arrangement.

Keep in mind the definition of refile: the destination country is unaware of, or disapproves of, the rerouted traffic. For instance in a BT document that is probably dated late 1999 or early 2000 it states: "The growing importance of refile is especially crucial. Any operator can obtain low prices for delivery to third destinations from US and EU refilers. There is virtually no destination country that does not have correspondent relations with at least ten carriers that will provide such services."

Refile providers to India from the UK, for example, include the following: France Telecom, World Exchange, Viatel, Optus, Mednet, Telecom Italia, Telia, MCI/Worldcom, TNZI, Telenor, and Sprint. Technically, this can be accomplished either with, or without, the knowledge and involvement of the third party country.

If done with the third party involved, the process is simple, and can normally be done with existing facilities, switching hierarchies and billing / call processing systems. The only difference between this form of refile and legitimate transit traffic is the intent, and the concurrence of the destination, so the technical requirements are nearly identical. There is no new hardware or software needed, so the investment is minimal.

If accomplished without the third party approval, the technical situation becomes a lot more complex. Since the existing switching hierarchy can not be utilized directly, the refiler must supply its own switching hierarchy, and billing / call processing systems. Depending on the details of the connection, dedicated network facilities may also have to be provided. In this case, the hardware and software could be very expensive and take a considerable investment of time and money to get into place. If the third party country does not condone the arrangement, the entire scheme could be at risk.

Refile is a method of bypassing existing operating agreements, putting further downward pressure on accounting rates. Today, it is very easy to confuse refile with other offerings, since it is often done without a lot of fanfare. The fact that a carrier is actually offering a refile product is often obscured, intentionally or otherwise. It can be a very high quality product, if the SS7 and Quality of Service standards are maintained. In essence, it is a Tier 1 carrier product, repackaged and resold.

Once a rarely used tool to hammer a recalcitrant PTT into lowing its accounting rates, today refile is a real product, and represents another step in the evolution of competitive telecom services. And interestingly, it provides a competitive offering to standard international services without losing the quality of primary connections.

Chapter 10
Changes
on the Horizon

It was against this backdrop of emergence, growth and change that competitors begin to win their first struggles with the telecom establishment for the international telecommunications dollar. At about the same time the Internet began its meteoritic rise. It could not have happened at a more opportune time.

At first, the greatest promise of the Internet seemed to be as a marketing vehicle to find new customers. Since techniques like callback were illegal in one way or another, the usual mass media methods of promoting the new service were severely limited.

Little did anyone realize that the Internet itself would ultimately provide a stealthy means of transporting gray market telecommunications services. The biggest impediment in installing private lines for international connections was that installation, and thus approval, had to be arranged on both ends of the circuit. Internet bandwidth can be arranged, in virtually every country, without such formal protocol.

Whether the "free" bandwidth available on the Internet is of sufficient quality to make this a viable alternative or not remains to be seen. We will discuss these issues in a subsequent chapter, but it goes without saying that the Internet has had a dramatic impact on the rollout of competitive telecom services in the new telecom world order. The Internet has been and will be used for everything from the aforementioned marketing to customer service, callback triggering, billing and much more.

Without the Internet, competitive telecom services would have probably occurred anyway, but undoubtedly at a much slower pace than the frenetic pace we see today.

Section II

The Impact of The Internet

Chapter 11
The Emergence of the Internet

There is no doubt that the Internet has affected the world of telecommunications. Its impact on telecom is even more dramatic than on nearly all the other industries combined. In retrospect, the rise of the Internet was completely logical and obvious, even as far back as the 70s. Everything was moving towards a greater use of networking as a major computer strategy. When Data General and DEC first proposed networks of computers as an alternative architecture to the IBM world of "big hummer" mainframes and "intelligent terminals," it appeared to be a stupid idea. But, in reality, the only thing missing was the bandwidth necessary for truly distributed applications. In 1978, a 9600 baud data line was considered "high speed". AT&T had just introduced the 212A modem, which offered 1200 baud over switched connections, and the Dataphone II allowed 9600 baud connections.

I remember working for the Bell System in 1976 when IBM ordered a 56K circuit from their manufacturing facility in Boca Raton, Florida to their headquarters in White Plains, New York. It took months to get that circuit up and running. Even then, it was very unreliable, and took a staff of technicians on both ends to keep it working. We could not imagine why anyone would need a 56K circuit — it was over 5000 characters per second! Obviously, no one could type at that speed. Big disk drives for 360 mainframes held about 10 megabytes of data, and at 56K, you could download an entire drive in under an hour. Why?

But the attraction of networking was contagious. IBM even had a strategy for interconnecting mainframes (nothing smaller needed to be networked) called "Systems Network Architecture," or SNA. SNA supported a variety of nodes, from the "intelligent" 3270 terminal clusters to the 360/370 mainframes. The PC was not even a glimmer in Bill Gates' eye yet. Undaunted by IBM, companies like DEC and Data General began to implement their networked architecture, aided and abetted by AT&T with its UNIX operating system, which provided a rather incestuous software solution to networking. It was very heretical, at least from the IBM philosophy of only networking mainframes.

At the same time, some people with no skin in the game, like the US government and a group of early geeks, in the form of scientists and academicians, began to adapt UNIX and some of the available hardware to work with the limited bandwidth then available.

Without tracing *ad nauseum* the development of the Internet, suffice it to say that even IBM unknowingly contributed to the growth of the Internet. IBM's launch of the Personal Computer (IBM 5150) in 1981 set the industry standard for personal computing. The unwashed masses now had enough power to dream about the possibility that some day they might be able to network their computers, and share applications and data.

The genie had been let out of the bottle, and by the mid-80s, while ARPAnet (the Internet's progenitor) was growing at unprecedented rates, dial-up networks, like the UUCP (acronym for Unix-to-Unix Copy Program) and bulletin boards were also proliferating. Although it is unclear exactly when Al Gore invented the Internet, the World Wide Web was emerging and started the whole revolution, but ... (for those of you who may not be in on the joke. US Vice President Al Gore created quite a stir when he tried to take credit for inventing the Internet!).

Regardless of who actually invented the Internet or the World Wide Web, it is quite evident that visionaries saw the tremendous opportunity offered by a ubiquitous communications media. Telecommunications visionaries saw uses for this new communications tool ranging from promoting their services to the more ambitious use of the Internet to actually transport the intelligence of a call. It is clear that the Internet has the capability to send binary data from one place to another, and it is a relatively easy task to convert voice to binary data. What is difficult, given the bandwidth available on the Internet, is to get the voice from one end to the other — consistently, reliably and with a level of quality that is satisfying to the typical telephone caller.

In the next chapter, we will look at the issues of IP Telephony vs. Internet Telephony. Why the first is a viable strategy, but the other is ham radio, and more work needs to be done before the latter will become anything more than the curiosity that it is today.

But there is no doubt as to the power of the Internet to support telecommunications in myriad ways — from finding and enrolling customers to providing back room support for telecommunications companies. Listed within this chapter are a few of the more intriguing applications for the Internet in telecommunications.

Pre-Sales Support

An obvious use of the Internet within the telecommunications community is to find, inform, sell and even close potential customers. The Internet has proven to be a valuable tool for all of these activities, but especially for gray market services, like callback and perhaps Internet Telephony or VoIP. One of the difficulties in developing customers for such services is that they are often frowned upon, discouraged or considered downright illegal. It is almost impossible to use standard mass marketing and distribution strategies to sell services that are not officially sanctioned. Most mass marketing advertising media will not, or cannot, accept advertising from gray market vendors.

Here is where the Internet shines; it has the ability to provide universal coverage, regardless of national boundaries. It helps to overcome the disadvantage of not being able to use mass marketing to sell services in countries where alternate services are considered to be gray (not Elisha) market. Because the Internet is universally available around the world, all kinds of services, whether locally condoned or not, can be offered, sold and serviced over the Internet.

If you are not convinced of the value of the Internet in selling telecom services, just go to Google (www.google.com), or your favorite search engine, and type in something like "telephone services", "long distance" or simply, "telecom", and stand back. Your PC will literally smoke as the plethora of service providers begin scrolling across your screen.

There has never been a vehicle like the Internet, and for emerging services, like competitive telecom, the Internet is like a mega-dose of steroids.

Customer Service

There are many ways in which accessible and effective customer service can be provided over the Internet, from billing detail, to remittance of payment. Pre-Internet, such customer service issues were major operational headaches. Providing these services over the Internet brings down many barriers to servicing a global customer base — financial, language, time zone and so forth.

Real-time, robust applications now exist, which use the Internet as their medium to give providers of alternative telecom services access to levels of customer service that were not possible pre-Internet. Because multilingual web sites can be arranged without a great deal of difficulty, services can now be provided in the user's native tongue, widening the appeal of the services offered.

Customers can now review billing detail, add or enhance services, pay bills, request service refunds, and change service levels, all from their computer, at any hour of the day or night.

Technical Support

Likewise, customers in far-flung locations have access via the Internet to complete suites of technical support offerings. Technical support has always been closely related to customer service, and they are often lumped together into a common delivery strategy. But technical support is really a disciple unto its own, and should be considered a stand alone business function.

Enter the Internet.

The Internet offers an excellent medium for providing a complete suite of technical support offerings, including troubleshooting, user guides, step-by-step operating instructions and even downloading of software upgrades and new features. Internet-based technical support can also host user forums where typical consumers can offer each other suggestions and support.

Agent and Rep Support

Moreover, the Internet can be used to great advantage to not only find and recruit sales agents and representatives, but also to provide the specialized support needed for these employees to do an effective job.

Sales agents can review and enter sales activity information, track customer activity and, with the right customer relationship management package, they can even identify customer needs and sales opportunities using operational data that's been gathered over time.

Service Delivery

One of the most promising, but also one of the most controversial, features of the Internet is its ability to carry voice traffic. A few years ago, a few of the Internet pundits were predicting that eventually all telephone traffic (worldwide) would be carried over the Internet. Of course, this was based on a very optimistic view that the Internet could "do anything". At the time the prediction was made, the total capacity of the Internet would only have been able to carry a small fraction of the world's voice telephone traffic and this assumes that the entire capacity of the Internet is brought to bear on this one application.

Without going over the mathematics involved, it should be obvious that the primary carriers of all telecommunication traffic (regardless of its nature) use the same physical facilities. Since the primary carriers and PTTs built these facilities, it is clear that they will not willingly surrender all of their capacity for low margin Internet traffic, i.e. let their high margin voice telephone traffic be cannibalized by unwanted competition.

There are also some serious doubts about the ability of current routers and data switches. Can they handle the relentless volumes of traffic that a public switched network carries every single day? In many cases the same routers and certainly, the same network facilities, are used for all voice and data traffic. So, the day that the Internet replaces the existing public switched network is probably still far in the future, but that's not to say selected applications can't use the Internet as an alternative transport mechanism. As an example, retrieving voice and/or fax messages over the Internet is very feasible and done quite often.

Enhanced Services

The Internet has vastly improved the offering of enhanced services. In addition to retrieval of certain messages, there are hundreds of applications that use a combination of Internet and traditional telephony to deliver their services. The classic example is Unified Messaging, which is discussed later.

What the Internet offers is an easy interface for users to do things that would be difficult, if not impossible, to do using conventional touchtone or voice recognition techniques. Imagine, if you will, that you want to use Speed Dialing for your calling card. You can program all the numbers that you want using the touchtone dial pad on your telephone. But, if you have a large quantity of numbers to program, the process can take considerable time. Now, let's say that the calling card provider has a switch problem and loses all your speed dialing numbers. You have to do it all over again, and if you programmed the numbers as you made the calls (as many people do), you probably don't have a list of all the numbers involved.

If you have an Internet interface to the speed dialing, the entire process might only take seconds to accomplish. This is just one example of how adding a web interface can make an existing application much easier to use, thus more attractive to users.

Chapter 12
VoIP

"It's programmable, it can do anything."

How many times have you heard this? Software designers, computer manufacturers and systems integrators have long chanted this mantra to consumers who are inclined to believe it, even if it's not true. I remember one example in particular. It was 1974, when AT&T introduced the Dimension PBX telephone switch, a programmable private telephone system designed to handle the voice telecommunications needs of medium to moderately large organizations.

This switch was the first truly programmable telephone system that the Bell System offered to its customers that was software driven, meaning it could be programmed on-site to perform certain functions. Notable to this discussion was the promise, oft repeated, that it would be the last telephone switch that a customer would ever need to buy — it was programmable.

As we have since learned, both in the voice telecommunications world and the computer world, even programmability has its limitations. PCs taught us that system designers and programmers will stay just ahead of the ability of the hardware to deliver, thus creating the first real practical barrier to programmability. Although in 1974, the world of applications that the Dimension PBX opened up appeared nearly infinite.

With this caveat in mind, we can now say that VoIP-powered telecommunications networks are truly programmable, and CAN do anything. (This is the first test of your recently acquired ability to recognize that there are probably practical limits to what can be done!) As with the Dimension PBX switch, there are practical limits that restrict the creation of new applications for VoIP, but the range is so great (at least from our perspective today) that we could almost say they are infinite.

In every telecommunications system, no matter how simple it is, there are three essential elements — the transmission medium, the intelligence itself, and the program control structure. For a simple example, let's look at a call from New York to San Francisco. The transmission medium is the collection of circuits and transmission paths that allow the electronic signal that represents the voices on the call to be sent from one end of the call to the other and vice versa.

The intelligence itself can mean the words and concepts that are carried over the transmission path. "Hello, Mom. How are you today?" is an example of the intelligence of the call. But, it could also be the electrical signal that represents a fax transmission, or another electrical signal that represents a modem connection.

Finally, we have the program that controls the flow of the call. For example, the telephone number that we dial to get the person we want on the other end of the call. There are also several layers of very sophisticated programs that arrange and manage the communication path itself. The programming is nearly transparent, and perhaps even magical to the user. That is, until something goes awry and the call drops or does something bizarre!

The more basic and thus the lower the functionality of the call, the more these three elements are isolated and operate independent of each other. For example, in a traditional call, the transmission medium is totally unaware of the intelligence that is being carried over the medium itself. Once the call is established, you can hit all the digits you want, but nothing will happen. Nothing will change; the other two elements of the system are totally unaware of it.

But, if we look at a more sophisticated system, like a voice mail system, we can see that the overall program control is reacting to the intelligence being transmitted over the communications link. As we interact with the system, we hear different messages, and can even program some of the features by the intelligence we transmit over the link. Let's say we type in the touchtone digits necessary to change the number of rings before the system answers from 4 to 2. Now we have changed the programming of the system using intelligence transmitted over the link.

(Please hang in with me on this seemingly abstract discussion for just a bit longer. We are going to make an important point soon!)

This is a very important development. It is one thing to have, for example, a system that can send unanswered calls to someone else by programming the system. It is significantly more important, if you can allow the user to determine the telephone number to be called in the event the call is not answered, and maybe even allow the user to determine the number of rings before the call is forwarded.

The problem with this type of user programmability is the difficulty in arranging a man/machine dialog. They speak in entirely different languages, and don't even think alike. The answer to this dialog in the telecommunications world has been to have the human speak in touchtone digits, and the machine to read back computer files that are converted to electrical signals representing a human voice. Voice mail and interactive voice response systems are some examples of this technological answer to facilitating man/machine communications.

In traditional telephone systems, there is a rigid barrier between the programming of the system and the intelligence carried over the system. Part of the reason for this is security, part is due to the difficulty in man/machine communications, and a large part is due to the fact that the architecture of the telecommunications control system and the architecture of the transmission system are very different.

But in VoIP, the architecture of these systems is much closer, in fact, the systems are often identical, using essentially the same server platforms, the same network connections and the same control programs. This commonality raises the possibility of integrating the functionality of the three elements — the transmission media, the program control and the intelligence, leading to some very interesting possibilities, the first of which have already begun arriving on the scene.

Chapter 13
Impact of VoIP

VoIP threw open the doors to international telecom market competition. It took what was introduced by callback to the next level. Callback still relied on the use of outbound international rates to compete with the national carriers in different countries. VoIP changed this by allowing the foreign, i.e. US carrier, to use their own circuits in order to create the minutes.

This was a total transformation in the competition structure of telecommunications. Why was this such a big difference? Well, traditionally telephone rates between countries have been set between the national carriers in each country based on a series of negotiations to which cost was only a minuscule factor, if a factor at all. Since the exchange rates between the countries were not based upon cost, it is fair to assume that these rates were set according to what those national carriers would like to make on the minutes, of course, without the added complication of competition. The private line business between countries has been competitive since the early 1980s and there are a multitude of suppliers that will supply bandwidth. Once carriers began to create their own minutes, they could sell them based upon a cost formula, as opposed to doing it in a vacuum in a monopoly environment. Some studies have suggested that the exchange rates that were set between countries may have been as much as 10 to 20 times or more of the actual cost of providing those services. Obviously that gave room for people like VoIP carriers to leverage on those rates.

But why was VoIP a critical factor in this? Clearly, competitive telecommunications carriers could have set up channelized services between countries and provided the same minutes, possibly at even greater efficiencies. What was the crucial element in VoIP that allowed it to succeed where the others were subject to a morass of regulations, harassment and intimidation by the national carriers?

VoIP could easily be disguised as being Internet. The letter "I" in VoIP means Internet as in Internet Protocol, and many countries have taken a hands-off attitude towards anything regarding the Internet because it is viewed as a new and critical technology that needs to be relatively free of regulatory fiat. Because of the architecture of the Internet, it is very easy for a carrier to acquire bandwidth between two points and use it as they see fit. Although some bandwidth can be used, and usually is used, to carry public Internet traffic, as for the rest of the bandwidth...well, carriers have the ability to use it for their own purposes. One purpose could be carrying voice traffic a la VoIP.

Please note: There is nothing in this that insinuates that VoIP or packet switched calls are in any way inherently superior to circuit switched calls. There are many companies out there that try to give the incorrect impression that VoIP has thrived because of some type of technical superiority that it enjoys over traditional or legacy telephone connections. This is not the case. It is because of the regulatory aspects that allow it to "come in under the radar". In other words, it is much easier for someone to sneak traffic in and out of a country where it's otherwise prohibited or subject to regulation and taxation when it is carried in a pseudo-Internet, i.e. VoIP format, as opposed to a channelized voice-type format. Therefore, one cannot discount the increasing proliferation of transmissions over the Internet, which are carrying telephone calls between countries.

The volume of traffic being carried over the public Internet continues to grow exponentially. The Internet is used for transmission of the majority of the world's email, most of the world's World Wide Web functions and many business functions, such as real-time data sharing (airline reservations, inter- and intra-corporate dealings, and so on) are being transferred to the Internet. With new encryption and security schemes, even corporate networks are using the Internet, at least for a portion of their data transmissions. Many large companies, whose employees work out of their homes (known as virtual workers, telecommuters or teleworkers), connect back to the office using the public

Internet. To say that traffic on the Internet has exploded over the passed five years would be a gross understatement. It is a very busy place, indeed.

The sad truth is that the public Internet has barely been able to keep up with the ever-increasing demand for bandwidth. More and more sites, for example, are incorporating streaming audio and video and other bandwidth-hungry applications. But, at the same time, the lure of a technical Valhalla that allows transmission of broadband facilities (i.e. T-1, E-1) for "free" is a temptation that few are able to resist. Many small telecommunications carriers are now attempting to carry all or part of their traffic over the public Internet by connecting routers to the Internet in one country, and connecting VoIP to routers in another country, thus transferring international voice traffic.

The reliability of these schemes is very much in question. Anyone who has watched a streaming audio or video presentation on the Internet and has seen it jitter, halt or stop, with a little sign at the bottom of the screen that says, "network connection buffering," will understand the difficulties of using the Internet as free bandwidth. Just because it's possible to make a connection at a line speed of T-1 at one end and a line speed of T-1 at the other end doesn't mean that you are guaranteed T-1 bandwidth between the two locations. This would be analogous to having an on-ramp to Highway 405 in San Jose that allows you to enter the freeway at 90 miles an hour at any time of the day or night. Even though you may be going at 90 miles an hour on the ramp, the moment you hit 405, you will be subject to the same traffic congestion and stop and go traffic as everyone else. A fast on-ramp does not guarantee a dedicated lane on the highway. This is exactly what happens when you connect T-1s to the public Internet at either end. You have T-1 access to the Internet peering point or to the ISP but there is no guarantee, in fact it is almost an absolute certainty, that you will not have that kind of bandwidth between the locations.

Having said this, there are companies who are attempting to, or may actually be implementing these VoIP schemes using the public Internet. While it might actually work, the quality of service provided by such schemes will vary by destination. For example, at the current time there is a great deal of bandwidth available in the US and the UK, so the probability is much greater that you will have a good deal of bandwidth, possibly even enough bandwidth to carry dedicated traffic within and between these two countries. But this "free" bandwidth is not guaranteed, so reliability is an issue even in these high volume routes. However, in bandwidth poor countries (e.g. many third world countries) you may be able to get E-1 transmission speeds to the ISP, but it's an almost certainty that you will not get enough bandwidth to support any level of dedicated communications. The picture becomes even worse if the traffic is between international destinations.

Advantages of VoIP

In addition to the significant advantage that VoIP has in regard to offering a considerably lower price for a basic telephone call, there is also the tempting promise of enhanced services. The real promise of VoIP lies in its innate ability to converge the power of the traditional telephone network with the unlimited potential afforded by a global Internet and computing power. This may sound like a line from a vapid dotcom prospectus, but that is the reality of the next generation telecommunications company. A view shared by financiers, entrepreneurs, and more importantly, by the consumers and end users of the technology who will ultimately decide if it has been successful or not.

In the following chapters, we will look at some of the "killer" applications, both promised and real, that are part of the VoIP culture, but first, let's investigate why VoIP has spurred this great interest in applications, and why VoIP service providers are so hot to make these applications real.

Chapter 14
VoIP Clearinghouses

With the advent of IP-based telephony, a new paradigm in exchanging traffic emerged. These often small, entrepreneurial enterprises base their existence on the ability to find at least one correspondent in a viable destination and exchange traffic with them. In order to be able to complete calls to destinations other than the one or more correspondent locations, the clearinghouse has to connect to other carriers, either VoIP, legacy, or both. Typically, the smaller clearinghouses with only a few routes use either legacy carriers or a larger VoIP company to provide the destinations that are not available on their network.

Many VoIP clearinghouses try to operate as arbitrage carriers, buying and selling traffic on small margins to build volume. The pitfall to this is twofold. First, the often scant resources of the small clearinghouse are stretched to the limit and the arbitrage business begins to consume them, preventing the clearinghouse from implementing their real business plan — a VoIP business. Secondly, the risk in arbitrage is great and the margins small, so one small slip up can mean the end of the company. This is not to say that the arbitrage business should try to limit its exposure to bad debt at all costs. The cost of credit policies that are too conservative may inhibit the ability of the business to grow.

VoIP clearinghouses have a distinct resemblance to Least Cost Routing Carriers, in that the product branding is the clearinghouse brand. The clearinghouse customer is generally unaware of which routes are served by the clearinghouse and are "on net", and which are passed off to others for completion ("off net"). The responsibility for quality of service remains with the clearinghouse, which must have some way of enforcing a consistent quality of service across all routes.

The rise of the boutique carriers and the increased availability of low cost, high performance switching and routing equipment, together with a highly competitive environment for traditional carriers, has led to falling prices for domestic and particularly, international voice services. The consumer recognizes little differentiation in a vanilla (without enhancements) voice offering; thus, price becomes the sole determinant. This leads to the extremes of competition where any service and any supplier that can meet bare minimum quality standards are preferred, i.e. commodity competition.

Chapter 15
Commodity
Exchanges

Commodity pricing exerts heavy downward pressure on prices and margins. Yet, lower prices are what stimulate demand. A few years ago, a three minute call from the US to Argentina could cost over $10. Today, many carriers offer rates well under 20 cents per minute retail, often with only a one minute minimum call. Thus, where a call to Argentina would have been a very rare occurrence because of the cost; today, one can place a three minute call once a week for a entire year for what a single call would have cost just a few years ago.

As you can imagine, this has fundamentally changed the way telecom service providers, large and small alike, operate. This telecom tsunami has sent economic ripples throughout the world. When a service as basic as telecommunications undergoes a draconian change, its impact is felt far and wide. At the onset of the millennium (in 2000, or 2001, depending on your calendar numbering system), the changes had become so fundamental that even the unthinkable was on the tip of many an industry pundit's tongue — the mighty, indefatigable AT&T might be teetering on the brink of disaster.

What had happened was simple. For decades, people knowledgeable about telecom predicted, projected and even knew that voice services would not be enough to carry the day. But, the warnings went unheeded, and well financed telecom companies and PTTs, alike, continued to build capacity. This glut of capacity led to the building of traditional and VoIP routes at an unprecedented rate.

The proliferation of services available through the new breed of telecom carriers has led to a telecommunications "flea market", where buyer and seller come together for the purpose of doing business. You can look over the goods, and select the supplier of your choice, and the product you want. The exchange extracts a small fee per transaction, or per minute for arranging the transaction. Buyers and sellers are free to negotiate the unit price between themselves.

There are several varieties of commodity exchanges, each with a slightly different orientation or target audience. All have the same underlying concept: as minutes and bandwidth become more standardized and the unit price drops, they become commodities that can be freely exchanged, much like stock shares or sugar or pork bellies. These exchanges may offer raw bandwidth in the form of private lines, or terminations, and minutes of voice-grade connectivity, to various destinations.

The architecture and operation of a commodity is much like a clearinghouse. In fact, the exchange may offer some standard clearinghouse services, like billing and routing. The key differentiating factor between a clearinghouse and a commodity exchange lies in the way the services are offered. Clearinghouses tend to have a set schedule of rates and each participant buys from and sells to the clearinghouse, independent of how the other end of the call is handled. In other words, there are no choices of carriers for a given destination. The clearinghouse makes the selection, similar to a stock exchange, e.g. you may buy a share of Exxon, but the stock exchange decides whom you buy it from. The clearinghouse does all the buying and selling, and adds a markup on each unit sold.

Telecom commodity exchanges are different in that they operate more like a true commodity market. In fact, some of the larger exchanges — Enron (US) and Band-X (UK) — were started by seasoned commodity brokers, who patterned the telecom exchanges in the likeness of a traditional commodity exchange. Arbinet was founded by people with their experience mostly in telephony, and thus has a slightly different orientation than the exchanges that were founded by commodity traders. For example, Arbinet has more emphasis on trading traditional bandwidth — circuit switched minutes and private lines — than Enron, who is almost exclusively oriented to IP trading.

These telecom commodity exchanges resemble the operation of traditional commodity exchanges, facilitating the matching of buyers and sellers. Buyers and sellers arrange a physical connection to the exchange, and then use a web-based tool to affect their transactions. In most of the telecom commodity exchanges, the buyers and sellers remain anonymous. All billings are payable to and from the exchange. In most cases, the transactions are relatively immediate, meaning that the services can be used very quickly after the transaction is complete.

The concept is so novel and radical to the telecom industry that it is difficult to predict to what extent it will be accepted. There are many factors, some as yet unseen, which may ultimately determine the success, or failure, of commodity exchanges. We should know the outcome in just a few years. It is likely that we will have one of two things to say about telecom commodity exchanges. Either, "Why, I remember when these things first started, now they are the ONLY way for carriers to operate", or "Remember the commodity exchanges that were around a few years ago?"

The line between traditional least cost routing carriers, VoIP clearinghouses and these new commodity exchanges is very hazy at best, and as they grow and broaden their offerings, it is likely that they will overlap their offerings, making differentiation even more difficult. It is hard to draw any conclusions as to what the real differences will be, but we will try to point out the existing differences, and the similarities.

It is interesting to note that with all these different models — carriers, clearinghouses and commodity exchanges — the basic underlying commodity is the bandwidth itself, and there are very few true bandwidth "manufacturers" in the world. It is an enormously expensive and complex operation to lay and maintain a fiber optic cable, or to put a satellite in orbit. It requires a multi-billion dollar investment in the facility, the equipment and the operational staff to keep it going. This is why most international fiber projects tend to be a consortium of carriers. Even the largest — AT&T, MCI Worldcom, BT, KDD and Telefonica — find it a strain to do a project of this magnitude on their own.

Given this scenario, the actual bandwidth, i.e. the commodity that the clearinghouses or exchanges are moving is probably the same bandwidth. This adds to already considerable irony — the actual bits that comprise the minutes sold by premium carriers like AT&T may travel the exact same fiber optic cable as those sold by the worst quality, cheapest carrier imaginable. These exchanges just serve to underscore those common routes.

Derivatives Trading

Today's telecom commodity exchange is a "spot market" exchange, meaning that all trades are immediate, or relatively so. Minutes or bandwidth are purchased on the exchange for consumption within a short period of time, with 30-45 days being the usual maximum period of time that the agreed upon rate is valid. In other words, the product is consumed almost in real time.

Traditional commodity exchanges have developed the notion of "derivatives", meaning that what is being bought or sold is not the commodity itself, but a future position in that commodity. If you go to the Chicago Merchantile exchange and buy hog bellies, you are not going to take them home and put them in your freezer until you are ready to sell them. You are really buying the right to purchase the bellies at a specific point of time in the future. When you sell these futures, you make or lose money on them, depending on the spot market at that time. It is almost like placing a bet. If the market for hog bellies goes up, you win, and if it goes down, you lose.

Commoditization of Voice Services

The concept of a telecom commodity exchange is highly dependent on the notion that telecom bandwidth is a standard commodity that can be exchanged as freely as barrels of oil. Complicating this is the sheer number of combinations that are possible in the telecom world. Over 500,000 unique different combinations are possible, with the number growing daily. In reality, today's telecom services and bandwidth probably are not as interchangeable as other commodities, although they are close. With a bit of work and some cooperation, which is not universally offered, the day could dawn where the majority of telecom services can be offered as freely interchangeable commodities.

Technical Standards

Key to the commodization of telecom services is the existence of an effective technical standard, which allows the various platforms, circuit and packet switched, to communicate with each other and to be able to pass calls, track billing and handle other administrative tasks necessary for telecom. This is a very thorny issue, since these standards have never been fully worked out for legacy circuit switched calls, much less for the newer, packet modes. Anyone who has ever tried to work between E-1 and T-1 trunks has seen the many technical pitfalls that can, and do, exist. If anything, the problem is more stringent with packet calls. The cumbersome implementations necessary for standards like H.323, which effectively allow open internetworking, gives one pause when considering the daunting task of interconnecting legacy circuit switched networks and next gen packet networks.

Part of the problem, if not the entire problem, lies in the age-old standardization battle. In summary it goes like this:

If you are a contender for any technical market, you strongly advocate the opening of proprietary protocols. It is the only way you can easily communicate with the majority of the systems already deployed. On the other hand, if you are the market leader, you want to keep the potential for networking at a minimum, since opening "your" market to competitors can only lead to a loss of market share and an erosion of your dominate market share position.

Today, Clarent has the dominant market position in VoIP gateways. This is due in part to its relationships with the "gorillas" of the market, like AT&T, BT and others. It is understandably reticent to open up any avenues that could cause its dominance to be challenged. There are many other manufacturers, who make gateways that are H.323 compatible, but are unable to fully interconnect with Clarent. These manufacturers are very anxious to make the supposed standard THE standard.

But, the ability to interconnect is essential to the telecom commodity exchange concept, so the users will put enormous pressure on Clarent to make its products more interoperable with the products of others. Over time, if the telecom commodity exchange concept does become the standard (amazing how many ways we can use the word "standard", isn't it?), then Clarent will be forced to fully support the technical standards.

This ends this lesson in Competitive Markets 101. Stay tuned for the next lesson, coming soon to a VoIP market near you.

Standardized Contracts

Also essential to the successful deployment of telecom commodity exchanges is the institution of standardized contracts. In the world of circuit switching and bilateral agreements, nearly every contract is individually negotiated. It is the way that things are done for many reasons:

1. It has always been done this way, so it fits the operational needs and methods of the existing incumbent telcos.
2. In an ITU promulgated world of bilateral agreements, individual contracts are easy to do, and can be tailored to fit the needs of each arrangement.

3. These bilateral arrangements often include significant investments and capital construction projects, and the agreement must reflect this.
4. It helps to keep the lawyers employed. You may think that this is a joke, but legal counsel typically has great influence with the decision makers, and will often prevail upon them to keep the existing structure in place.

If telecom commodity exchanges are to succeed to any measurable extent, a multilateral, standardized contractual arrangement must be established, or some of the most significant advantages of telecom commodity exchanges will not be realized. There is a good bit of progress in this area, and some standardized contracts are starting to emerge. This is an area that will undoubtedly evolve significantly over time.

Quality of Service

Unfortunately, telecom services and bandwidth are not a "one size fits all" commodity. There are many different applications for telecom bandwidth, and even switched minutes have a wide range of quality that can be required. For example, the requirements for a "really cheap" prepaid card, whose purpose is to say "hi" to Mom once a week, is totally different from a fax machine. The prepaid voice application can tolerate less quality in the voice signal, the fax machine cannot. At the same time, unlike the fax customers, prepaid customers are much less appreciative of long call setup times and intolerant of frequent busy signals.

One recent startup company claims to be able to offer over 256 different levels of quality. It is unclear as to how many are really needed in the marketplace, but the ability to provide a consistent and predictable quality of service across numerous destinations is important, and a real challenge for telecom commodity exchanges to provide.

Universal Access

Universal access is the holy grail of telecom commodity exchanges. Their success and survival is probably more dependent on broadening the availability of their services across a wide range of physical and logical connection types than any other single criteria. The success of the Internet was due, in large measure, to the great availability afforded by peering points, like MAE East and MAE West. Telecom commodity exchanges have the same challenge — make your services available in as many formats and as many physical meeting points as possible.

Telecom commodity exchanges which emphasize both packet (IP) and legacy circuit switched protocols have even a greater challenge, since the meeting points for Internet/packet/IP connections are not the same as those for legacy carriers. This is probably one reason why most of the current exchanges specialize in one method or the other. Supporting both doubles much of their overhead, both in technology and network.

Network Requirement

The next obstacle facing a telecom commodity exchange is the network required to support it. By definition, a telecom commodity exchange is a locus of activity. This means that the network backbone of the exchange must be wideband, fault tolerant and have the management tools required to support it.

Anonymity

Anonymity of the participants is a dual edged sword. On one hand, the very fact of anonymity allows a certain promiscuity on the part of those involved in the exchange. On the other hand, not knowing what company you are exchanging traffic with prevents long term relationships from developing. The relationship is with the exchange, and the relationship of the participants is between themselves and the exchange, not between each other.

ADVANTAGES

✔ Anonymity

✔ Time to Market

✔ Carrier Neutral

✔ Market Facilitation

✔ Credit Risk Avoidance

✔ Variety of Choice

Figure 6. Advantages of Exchanges.

Section III

Interoperability

Chapter 16
Switching Platforms

Not too many years ago, there were basically two types of telephony switches available, the Central Office and the PBX. The range of choice in features, prices and manufacturers was very limited and specialized switching needs were rarely addressed, if at all. The increasing demand for telecommunications has created the need for a wide range of specialized applications, including callback, 1+, international resale, cellular, paging, ACD, CLEC and many others.

At the same time, technical advances in switching technology and control systems have made it possible to address those needs, and produce switching platforms for highly specialized and demanding applications. These new products are often produced by a new breed of switch manufacturer and vendor tuned to providing the needs of very specific markets. Developments in CTI (Computer Telephony Integration) have made it possible to intelligently link computer systems, Interactive Voice Response (IVR) systems and telephone systems together to create flexible and reliable platforms.

VoIP introduces a new technological twist to voice communications — packet switching. Not that packet voice is new itself, it has been around for awhile, but the complete package offered by VoIP — packet switching, Internet compatibility, and ubiquitous access — provides an opportunity similar to that originally offered by callback, but multiplied many times over. Later, we will see why this opportunity is so much greater than callback. But first, a quick look at the technology.

You will find that the basic concepts, many of which date back to the invention of the telephone in 1876, are still minimum requirements today, so it would be best if you read this entire section. We will try to make it as interesting as possible.

Chapter 17
Legacy Systems

What a difference a few years makes!

Manual switchboard technology dates back nearly to the invention of the telephone. These switchboards were the workhorse of the industry and they
remained firmly entrenched from the beginning until the first automated systems began to be deployed in the early 1950s. It has been less than 15 years
since the last manual switchboard system was taken out of service in the US.

It is tempting to just skip over the whole area of legacy systems and, like so
many of today's technological gurus, simply write off the entire installed base
of telecommunications systems prior to the introduction of computer-based
packet systems. We could then just focus on the new exciting area of VoIP and
completely ignore the past! The only problem is that the vast majority of the
world's voice telecom traffic today is carried by legacy systems.

Even if you decide to jump over the old, much maligned legacy systems, and
deploy only VoIP technology, you will eventually have to interface with the
existing telephone network somewhere, sometime. So, even you mavens of
modern technology are going to have to face the roots of telephony for several
years to come.

Sorry!

Even if you grant (which many do not) that IP is inherently better or more
desirable than circuit switching, you still have to deal with the reality of legacy
systems — they will be around for a long time to come. In order to understand

how VoIP works, it's essential to have a solid understanding of the basics of traditional telephony. Although the real promise of VoIP is its ability to significantly enhance traditional telephony service, there is one minimum requirement that all VoIP must meet — integration with legacy systems of all varieties.

Let's start with the first telephone system, as invented by Alexander Graham Bell in 1876. You recall the story of how he placed a transmitting instrument in one room, and a receiving instrument in the attic, a few feet away, and his assistant, Thomas Watson heard his voice. The transmitting instrument was able to create an electric current that was modulated by the physical vibrations of Bell's voice. The receiving instrument converted the modulated current back into physical vibrations. This is still the basic principle in today's telephony. All of the enhancements have been in the network itself!

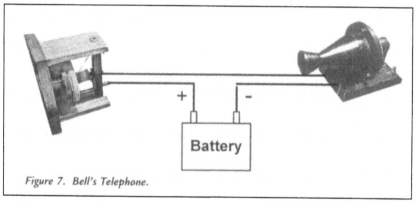

Figure 7. Bell's Telephone.

The purpose of the battery is to create a flow of current through the instruments. This device is very limited; in fact, it can only handle a "half duplex" transmission. The circuit can handle only one conversation in one direction. Presumably, Watson had nothing to say back to Bell!

In order to make this device "full duplex" so that both parties could talk and be heard by the other party, the instrument needed to be both a transmitter and a receiver. This still holds true today. A headset has a transmitter (the microphone) and a receiver (the speaker). Of course, built into today's telephone is circuitry that reduces the echo or "talkback" that a user might hear, especially if two telephones are located some distance apart. Try using Microsoft's NetMeeting application to see how disturbing this effect can be. This duplexing of the receiving circuit with the transmitting circuit is important because having a separate circuit would have doubled the number of wires needed to connect two points.

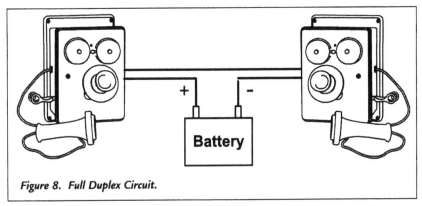

Figure 8. Full Duplex Circuit.

The next step is to allow one of the parties to signal the other that they want to talk. In a simple "hoot and holler" intercom system, such signaling isn't needed because everyone is listening and can "hoot" or "holler" if they want to be heard. But, a more genteel solution is to have a ringing device on the

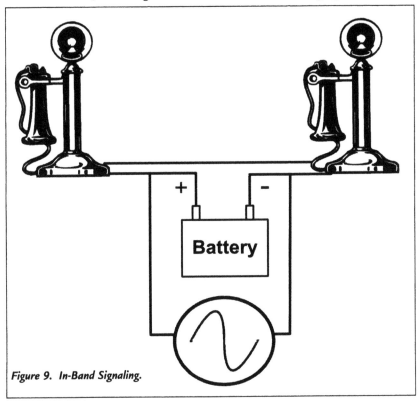

Figure 9. In-Band Signaling.

telephone that allows one party to signal by sounding a bell, buzzer, or other signaling device on the other instrument.

In the case of legacy telephones, this traditionally took the form of an "in-band" signal, so called because the signaling trigger shared the same circuit used to carry the voice signal. Again, this allowed a single pair of wires to be used to connect two telephones. Had these innovations not been created, it would have taken three pairs of wires instead of two to provide telephone service.

Now, we have a pretty functional, point-to-point telephone system, that will allow either party to signal the other that they want to talk, and for each party to talk and to hear and be heard by the other party.

But, these two people can only talk to each other, and only when both are situated very close to the instrument. As practical as this device may be, a HUGE improvement would be to allow "many-to-many" calls, in other words, anyone on the network could pick up an instrument and signal, and converse with, a person at another instrument. This development moved the telephone from a technological curiosity with limited application to one of the most important tools of the modern age.

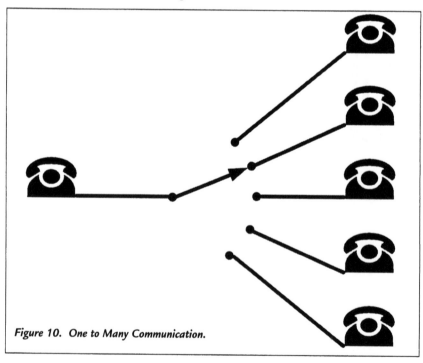

Figure 10. One to Many Communication.

Let's start with a "one-to-many" structure that will allow a single operator to initiate and converse with multiple parties. To be useful, the operator must be able to signal and converse with each party independently.

The simplest way to accomplish this is to bring each pair of wires from each user to a separate jack on a patch panel. The operator can then plug their instrument into the user's jack and signal and talk to them. Allowing each user to independently signal the operator is accomplished by giving each user a ring generator to ring a common bell at the operator's location, and then to place a light above the jack for each user, which lights up to indicate which user is ringing the common bell.

Are you beginning to see that whenever you engineer a telecommunications circuit, you have certain requirements that always have to be addressed? Talk paths, signaling, control — these are the fundamentals of any communications circuit, no matter how sophisticated they are. This includes VoIP.

If you are still curious and want more details, just visit the appendix on TCP/IP for a complete discussion. As long as you always remember to keep the three elements (talk paths, signaling, control) in mind, and you will understand most of the requirements of almost any network.

Bell had the insight to realize that the telephone network needed to be more

Figure 11. Switchboard Operation.

than just a point-to-point intercom arrangement, and that the real power lay in being able to summon a particular person or location on demand. He and Theodore Vail envisioned a world made smaller by the ability to "reach out and touch someone" whenever a person wished. (One wonders if Bell would have unleashed his invention on the world if he had had the foresight to see telemarketers in the future!)

The next major step forward would be to allow the operator to take an incoming call from a user, place a call to another user, and connect them. After some attention to electrical details, we can use our patch panel system as a "switchboard" that can function as a local "central office", switching calls between users as required.

This operating architecture began in the late 1800s, and continued in the US, without a lot of changes, until the late 1950s. Manual switchboards are still in use today in some remote areas outside the US and in private networks. The operation is simple. Each user has a jack on the panel, and the operator has an array of cord pairs. The operator answers an incoming request by inserting one half of a cord pair into a ringing jack. After determining the destination of the call, the operator plugs the other half of the cord pair into that user's jack, rings that user, and once the user answers, the operator disconnects. Special circuitry informs the operator when either party has disconnected from the call, and thus both plugs can be removed.

Figure 12. Inter-CO Trunks.

This architecture allows a group of instruments that have their circuits centralized into a "central office" to communicate, but has some practical limitations, based on the number of users, and the distances involved. It is a simple fact that in a normal environment, a vast majority of calls are made to destinations that are geographically close to each other. But people from this group of users might want to call users in other central offices, so the architecture was enhanced to add "trunks" to other nearby central offices. These trunks were shared by the entire group of users, and required operators on both ends of the trunk to set up and disconnect the calls.

When the call was complete, the operator had to physically disconnect the two legs of the call, and write down a billing slip. (These were known as "toll tickets", which were so deeply ingrained in the Bell culture, that when I started working for Southern Bell in 1973, all employees still were required to certify that they had never destroyed a toll ticket. However, since they were not used after about 1968, I never even saw one!)

Now we have a way to connect two central offices, and can expand this to

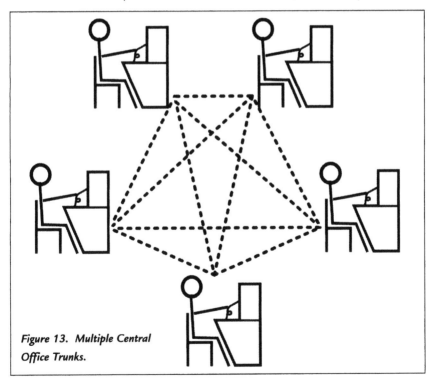

Figure 13. Multiple Central Office Trunks.

more, through the use of separate groups of trunks for each CO we wish to connect. Every CO would have to have a trunk group for every CO to which they wanted to be able to connect calls. They would use these trunks to connect their users to users in the other CO.

It was possible for a CO to act as an intermediary, i.e. arrange a connection from one CO to a different CO. This is called a "tandem" connection, and involves the participation of three operators to make the connection. These "long distance" calls were very labor intensive, making them limited in availability and expensive to set up and manage. Although through the use of such arrangements, long distance calls could hop-scotch from central office to central office, providing long distance service throughout a country. This complexity helped to justify the higher rates charged for calls that were increasingly farther from the caller's local central office.

It is interesting to note that the biologically growing VoIP network today has some of the characteristics of the first long distance networks, i.e. calls are

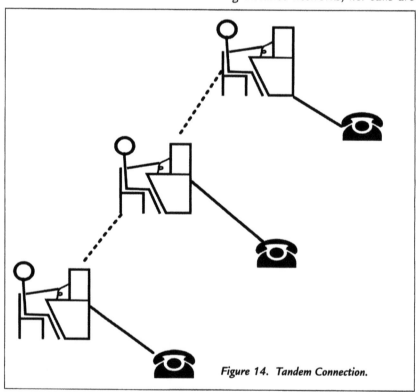

Figure 14. Tandem Connection.

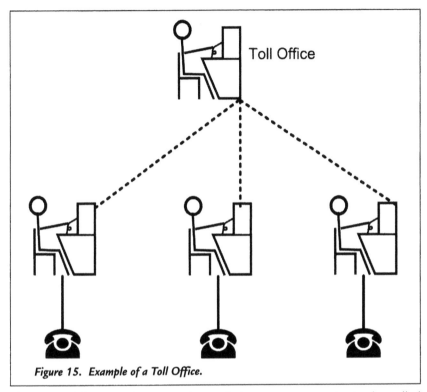

Figure 15. Example of a Toll Office.

often routed from carrier to carrier until they get to a carrier in the called party's local area.

As COs began to proliferate, and the traffic between offices increased, it became obvious that an architecture similar to a CO could be established to connect the calls between Central Offices. Sort of a "switchboard's switchboard," called the "toll office." Each local CO would have trunks to the toll office, which then arranged the connections between COs. A toll office would not have any end users of its own, and would serve only as a long distance carrier.

This naturally expands to include higher levels of regional offices, which are part of the so-called "Class" structure that traditional telephony types will often refer to. At the lowest end, are the local COs, which are called Class 5 offices, and on up to Class 1 regional toll centers. These offices are organized into a hierarchy, with higher volumes of traffic and bigger switches being used as you moved up from Class 5 to Class 1. This traditional architecture in the US has changed and flattened due to several factors:

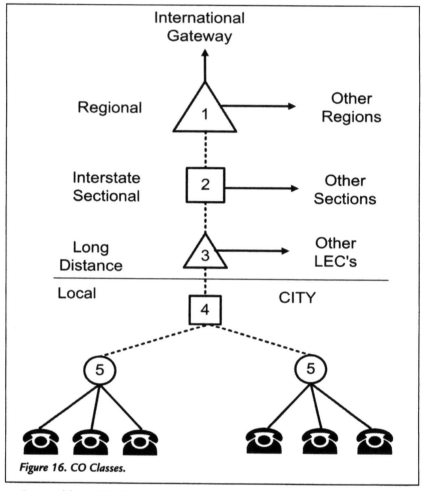

Figure 16. CO Classes.

Competition: AT&T's competitors have created their own hierarchy, thus the difficulty of handing calls off from one hierarchy to another have led to common meeting points.

Greater Bandwidth Facilities: One of the reasons for a hierarchy was to decentralize the switching due to limits in switching capacities. As fiber optics begins to dominate the US network, there is more capacity concentrated in one backbone.

Cost of Switching Systems: Switching systems today have traffic handling capacities that make the need to offload switches obsolete. It is now more economical to have a lesser number of switches, handling a greater level of traffic.

Until the advent of VoIP, circuit switched, center stage switches in a hierarchical network dominated as the predominant architecture for telephone service around the world. Many of the same structural elements of the legacy networks have also been incorporated into the VoIP structure, for the simple fact that VoIP is an emerging technology and it will have to co-exist with legacy systems for some time to come.

Automation Enters the Scene

Despite its technological shortcomings, the live operator type of call processing system was used throughout the Bell System, and other telephone compa-

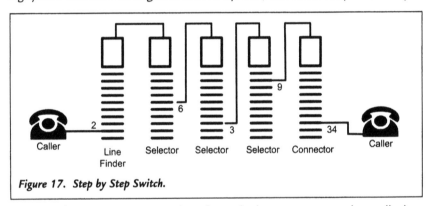

Figure 17. Step by Step Switch.

nies, into the 1950s. It may not have been the best way to complete calls, but it had some advantages. Features like operators who would continue to call busy numbers and connect them to the caller when the line became available, or even take messages for customers, are not available today, at any cost. In some small towns, the operators were like 411, information, yellow pages, Consumer Reports and even the local gossip column — all rolled into one. This was true customer service, though it was inefficient and labor intensive, and therefore expensive.

During the 1950s, Bell began replacing the manual switches with a series of electromechanical switches, called "step-by-step" or just "step" switches because of the way they stepped up and across a collection of mechanical contacts to make the connection. Each digit dialed was actually handled by a specific switch. If you went into a step office during the busy hour, the noise was overwhelming. Technicians wore earmuffs because of the high noise levels. At about the same time, Bell began using banks of relays to route and control the calls.

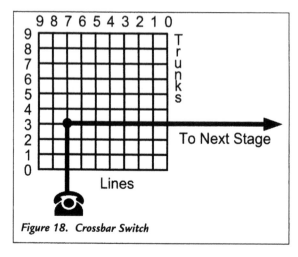

Figure 18. Crossbar Switch

The banks of relays evolved into a faster, more efficient switching matrix, called a "crossbar" switch, because of the grid-like design. Although much faster and more efficient than step technology, it was still mechanical, but much smaller and used considerably less energy. Bell also began using computer technology to handle some of the routing and management of the network, and in the 1960s, introduced "Direct Distance Dialing" (DDD). This innovation promised to automate the process of making long distance calls by eliminating the operators, allowing the user to complete their own calls simply by dialing numbers to make the connection. At the same time, developments in transmission technology allowed Bell to leverage its existing copper network to carry a much greater quantity of calls.

Through the 1960s and 1970s more and more computer and solid state technology were introduced by Bell, thanks to the innovations of Bell Labs (the place where solid state technology started with the invention of the transistor in 1947). By the end of the 1970s, a completely solid state switching system called the Electronic Switching System, or ESS, had been introduced. Even the switch matrix itself was computerized, and the first use of Time Division Multiplexing (TDM) began wide spread deployment.

Bell Labs was in overdrive, to the extent that with so many innovations rolling out, President Johnson once called Bell Labs a "national resource". Satellite transmission, digital radio (microwave) and compression and multiplexing technology gave the Bell System tremendous network capacity. It was dominant in computer technology innovation, despite the fact that it was barred from the commercial sale of any computer technology by virtue of a 1956 Consent Decree. This technology was limited to internal Bell use only, along with a few domestic and international customers, who used its network switching products.

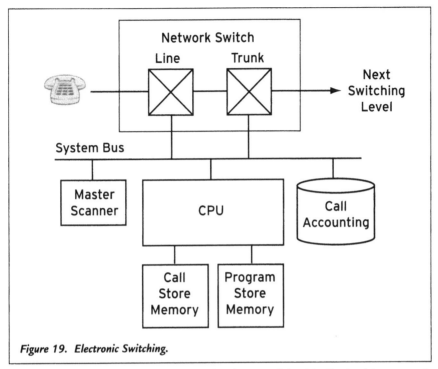

Figure 19. Electronic Switching.

Through the use of all of this technology, Bell had built the biggest and most powerful circuit switched network in the world. The rest of the world followed suit. Although a sometimes perplexing technological Tower of Babel has emerged, the world's PSTNs (Public Switched Telephone Networks) still operate on a call by call basis over a switched line, giving each call exclusive use of the facilities for the duration of the call.

Chapter 18
Packet Switching
vs. Circuit Switching

To understand all of the controversy, we need to examine the existing state of both packet switch and circuit switch technology.

Circuit Switching

The traditional telephone network is a circuit switched topology. This means that to create a communications channel between two (or more) parties, a channel has to be created end to end, and back again. These channels are usually full duplex, meaning that conversations can proceed simultaneously in both directions. Before the advent of digital techniques, a telephone circuit was a dedicated end-to-end channel whose purpose was to carry the signals necessary to conduct a telephone call.

When the cost and time involved in providing each party with its own pair of copper wires from end to end became too great, new technologies like Time Division Multiplexing (TDM) were created. Because the inherent bandwidth provided by a pair of copper wires far exceeded the needs of a single voice telephone channel, more than one call could be sent over a single pair of wires.

TDM uses a single bus, or wire, to allow multiple calls to share the same transmission path. These calls are gated by a sophisticated clocking system to give the path entirely to one call for a set period of time, then to the next call, and so on. This is called a "synchronous" system because of the need to accurately clock each call to a standard time clock.

While this allows the carriers to achieve a much higher capacity using their existing copper networks, it has a certain amount of inefficiency. Mainly because each call is given its time on the network path regardless of whether it has any intelligence on it or not. If 24 calls share a common path, each call gets 1/24 of the capacity whether it needs it or not (although modern systems can detect inactivity and "steal" bandwidth when possible).

Inefficiencies notwithstanding, TDM is a great innovation and it accelerated the rollout of the US domestic telephone network, as well as international connections. In the US, AT&T developed a standard for this multiplexing, called T1.5. This standard called for a 1.544 megahertz channel (well within the capabilities of a twisted pair of copper wires) capable of carrying 24 simultaneous telephone calls. This effectively multiplied the capacity of AT&T's fledgling network by a factor of 24, for an investment that was only a fraction of that.

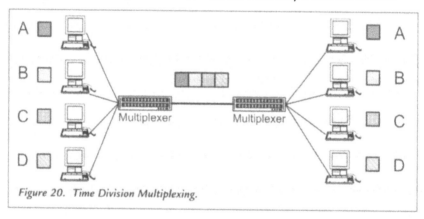

Figure 20. Time Division Multiplexing.

In Europe, a different standard, E-1, which put 32 channels on a single pair of wires was developed. Of the 32 channels, 2 are used for signaling and control, leaving 30 available to carry calls. This "standard" is not as "standard" as the T-1 standard, with many countries having their own standard. As explanation for this seemingly confusing and contradictory statement, suffice it to say that a T-1 in any country that uses it — North America and Japan, primarily — is relatively easy to connect from end to end.

E-1 circuits, while more pervasive around the world than T-1s, are not as standardized. Most European countries have their own E-1 standard, like the Swedish P6 that is used within that country. Some of these countries also support a standard, CEPT (Conférence Européenne des administrations des

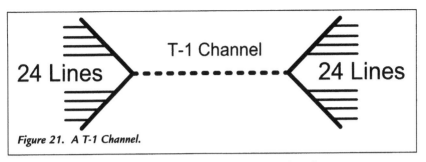

Figure 21. A T-1 Channel.

Postes et Télécommunications), which is an organization that attempts to pre-scribe some technical standards among European nations. But, there are always troubling details in connecting these E-1s — details that are significant enough to render even the best engineered facility completely unusable. For this reason, most carriers and resellers use ISDN for these trunks. ISDN, in addition to being much more rigidly engineered, and thus more compatible, also offers very convenient and useful Layer 2 information, which provides more data and information. Layer 2 is closely tied to SS7, making the process of connecting at a very high level much more feasible.

The advent of higher bandwidth transport media, like microwave ("digital radio") and fiber optic networks, are further enhancements of these technolo-gies, and can accommodate thousands of calls on a single circuit. But, nonetheless, users still expect, and receive a "dedicated" real time connection from one end to the other.

Think about it, once you call someone and they answer, even if neither of you say anything for one minute or so, you still have the use of the channel. (This may not be entirely accurate, since carriers can detect this silence, and "temporarily" allocate your unused bandwidth temporarily to other users who are more active.) There is an obvious waste and cost factor involved here, since the carrier must allocate bandwidth for each and every call, and maintain the connection for the users from the beginning of the call to the end.

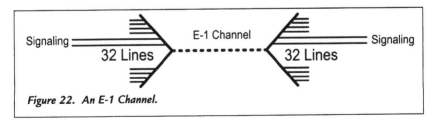

Figure 22. An E-1 Channel.

The other network element of circuit switching is the switch itself. A telephone switch is a major league piece of equipment. In fact, a telephone central office is basically a building designed to house and protect a switch. There are obviously more functions than just switching, but that is the primary function and justifies the investment in the rest of the building.

The Rise of VoIP

We now have enough background in the operation of a state-of-the-art, legacy, predominantly circuit switched Public Switched Telephone Network. Now we can begin to understand how VoIP works, and why many people think it's destined to take over all voice traffic.

One quick note, however, to make sure that there is no misunderstanding. Regardless of what you may read about the imminent demise of legacy switches and networks, it is *not* imminent, because of simple economic reality. There is an enormous investment in these legacy networks that makes it uneconomical to phase them out over a short period of time.

Remember the Polaroid instant camera? When it was introduced in the late 1950s, most people believed that Eastman Kodak, the dominant producer of film and film processing, would soon be out of business. In fact, fifty years later, it is Polaroid that is now facing extinction in the face of the digital camera.

What happened? It is an abject lesson in the economics of inertia. When Polaroid introduced its camera, there were thousands (maybe, millions) of cameras in the hands of professionals and amateurs alike, who were not going to throw away the investments they had in film-based cameras, even if the new Land cameras were superior (which, of course, they were not, and still aren't!).

There are a multitude of examples of how such investments in legacy equipment cause inertia which, in turn, give the existing "state-of-the-art" a stickiness that is hard to overcome. In the world of telecom, we have seen many examples of this inertia — touchtone, electronic key systems, digital switches, voice mail, ESS central offices, and so on. VoIP is no different, the speed at which the existing embedded base is replaced will be dependent upon several factors; most of which do not relate to the technical merits alone. In fact, the technical merits are often one of the least considerations in the equation — again, *economics* comes to the fore. How much was invested in the existing technology? When will it be amortized? What is the economic benefit to the newer technology?

The dominance of VoIP, or some packet oriented protocol, must be delayed while investments are amortized. Thus, a new best of breed packet oriented technology cannot be guaranteed success until critical mass is reached. If the rollout is too slow or stalls at any point there may be a newer technology that takes over before the conversion is complete. Look at satellite voice communications, it was destined to carry virtually *all* long distance traffic, but a slow rollout allowed fiber optics to gain ascendancy before satellite technology reached its full potential.

The same could happen to VoIP.

VoIP appears at times to be an eclectic amalgamation of the PSTN and the Internet. Yet, if VoIP had started in one or the other arenas, it might have developed differently. With the current mix, VoIP must permit internetworking between the legacy, circuit switched world, and the nex gen, packet oriented world. We will see that the gateway is a critical component, because it forms the interface between the PSTN and the Internet — between the old and the new.

A natural consequence of the computerization of the telephone network was the greater use of computer technology to actually switch calls. This trend continued as personal computers exploded on the market. New companies, like Dialogic, emerged and created hardware that could reside in a PC and allow it to interface with telephone networks. Simultaneously, the Internet began its stellar rise. In fairly short order, a great number of users had access to the "free" bandwidth available on the Internet. New hardware that added voice grade equipment to PCs also began to proliferate, and lights began clicking on in fertile minds everywhere.

Around the world, people begin to ask, "why can't I use the Internet for my telephone calls?" Simultaneously, the attention of the telecom industry, like most other businesses began to focus on the Internet. It was demonstrated that calls, given the right circumstances, could be routed over the Internet. The quality of the calls varied, and reliability suffered, but the concept took root and begin to grow.

(I wonder what the person who first made an Internet telephone call said? "Mr. Gates, come here, my PC just crashed!"?)

At the same time, competition for voice telecom services began to reduce unit prices for long distance. In the US, retail rates commonly fell 20% or more each year. This caused the demand to skyrocket (remember the Economics 101 Supply & Demand Curve?), so carriers sought ways to create capacity.

While all this was going on, the demand for data services, like packet oriented Internet transport was growing exponentially.

Carriers who owned the underlying facilities cashed in on the Internet rush by offering Internet access, backbone facilities, along with web hosting, design and implementation, etc., creating an even closer connection between the Internet and traditional telephony. The only big departure from traditional telephony was the increased use of packet technology for the transport of voice services.

Packet switching of calls introduced an "asynchronous" technique to the formula. In this technology, each end point (perhaps a caller) sends packets of data representing the digital encoding of the voice signal over a common circuit. Each packet has an embedded address, called a "header," that informs the network where the packet should go. While most networks limit the size of individual packets, the end points are free to send as many packets as they wish, whenever they wish. The packets are re-assembled on the receiving end and decoded back to the data stream from which they were encoded on the sending end. This creates what is termed a "virtual circuit."

The significance of packet architecture is that there is no theoretical limit to the number of packets that can be sent. Sound remarkable? Well, it is true, and not "voodoo technology." (We'll cover that later on!)

What's the hitch, you ask? There surely must be one.

Well, of course there is one slight detail. It is called "latency", which means that as network traffic builds, packets might experience a delay before arrival at the

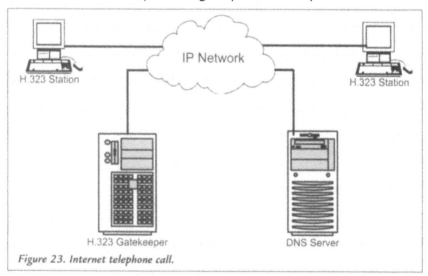

Figure 23. Internet telephone call.

end point. For an analogy of this, let's go back to our crowded highway that we used before. If you are on the on ramp, and there are no other cars in front of you, you can proceed at the maximum allowable speed to the highway itself. On the other hand, if you are in exactly the same position on the ramp, but there are 25 other cars in front of you, and each one has to merge separately onto the highway, it is going to take you much longer to get to the highway. This is latency.

Some latency is to be expected, and can be tolerated, depending on the type of information that is being sent. Text-based data can tolerate some fairly extensive delay, but voice-based data, which involves "real time" humans on each end, is much less tolerant of delay. In fact, the ITU established (based on a multilingual study a few years ago) that delays of over 1/4 of a second made the circuit difficult to communicate on, and a half second delay makes it virtually unusable as a voice circuit.

The "IP" in IP Telephony or VoIP refers to "Internet Protocol". It is the very same IP as in TCP/IP. A Protocol in this context is the pre-arranged sequence of events and signals used to allow a device on one end of a digital circuit to communicate with a device on the other end of the digital circuit. It facilitates the flow of information because the data flowing from one end to the other follows a rigid but terse set of instructions (that allow the information to flow).

The "Internet" in Internet Protocol is actually a lower case "i", or internet — a connection between networks, which is why it is called an "inter"net. This is as opposed to "the Internet" — the public internet that we all know and maybe love. Are you sufficiently confused yet? Hang on, we're almost there.

When all the devices that need to talk to one another are on the same network (or same circuit), the job of sending data from one end to another is relatively easy, since each device can easily be assigned its own address. Private networks do not have to worry about having addresses that are unique outside of their own boundaries, since theoretically, nothing will go outside to another network. If every possible device in the world had to have its own, universally unique address, we would quickly run out of addresses. It would also require that every private network in the world to register all of its devices individually, and would create a bureaucratic nightmare for organizations like ICANN, who administer these numbers.

To avoid this dilemma, the Internet Protocol was defined in such a way to allow "off net" addresses to be transmitted by way of a single gateway which is "on net". This device will route the data to other network end points, possibly in other private networks. In addition to allowing the efficient routing of packets, Internet Protocol also permits the originating end point device, to specify the nature of the content of the data, so the device(s) on the other end can handle it appropriately.

Now, if you remember the previous discussion on the PSTN, you will see that the Internet Protocol has many of the same features and capabilities as the PSTN. There are routers that operate like switches, network facilities, billing systems and interfaces to customer facilities. In fact, once you get under the hood of even a legacy telephony network, much of the hardware is exactly the same as the hardware that you see at an Internet Service Provider (ISP).

Chapter 19
VoIP Network Architecture

Let's take a look at some of the technological pieces that make up VoIP. We are first going to look at a "pure" VoIP scenario, and then we will investigate how it can integrate with existing telephone instruments.

To start with, we need a means to convert the human voice to an electrical signal, and a device to convert an electrical signal to sound that the human ear can hear. Well, with some updates, Mr. Bells invention still works, and in fact, although today's telephone doesn't resemble Bell's phone very much, it's functionally identical in its human interface.

Next, we have to arrange to digitize this voice, converting the analog electrical signal to a series of pulses. We will not go into detail on this process, since it is so well described elsewhere. What is critical to understand is that the encoding/decoding scheme that is used is essential to the successful transmission of the voice signal. The device that does this coding and decoding of the digital signal is called a codec, and can be hardware, software or a combination.

Obviously, both ends must use the same transport protocol, or the voice signal cannot be understood, but less obvious is that the signaling protocols must be understood by both ends, as well. We will cover the protocol issue in more detail later, when we talk about H.323.

For now, let's use the basics of Internet Telephony, but instead of moving the packets over the public Internet, with all of its delay and congestion issues, let's move them over a higher bandwidth, low latency network, like a LAN. The communications functions will be the same, but with the "guaranteed" bandwidth, the end result will be a usable voice communications link. One that real human beings can actually use, and one that will not have the nagging quality issues that go with "free" bandwidth.

But, that is the point of VoIP. We can use any existing data connection to transmit voice, with some caveats. We can even use the Internet, if it offers enough bandwidth, and a low enough propagation delay.

The actual network elements of VoIP are essentially the same as they are for TCP/IP, and the Internet, so we will not cover them in detail here. We will instead concentrate our attention on the endpoints. This is where the real action is as far as arranging a VoIP network is involved. Please keep in mind that no amount of hardware or software can overcome the problems of a low quality circuit. We will cover this in greater detail under Voodoo Technology, but be forewarned, if the network is not adequate for the quality and bandwidth requirements of the service you intend to send across it, you (and your customers) will be very unhappy with the results.

The first full scale VoIP systems used proprietary signaling protocols, mak-

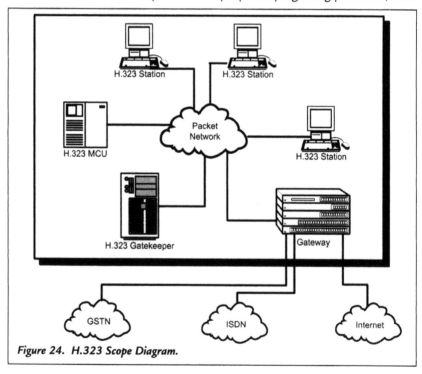

Figure 24. H.323 Scope Diagram.

ing them incompatible with each other. This was no mere oversight — telecom manufacturers are notorious for making sure that their products will only talk to their products (in particular the manufacturers who have their sights set on market dominance). That way, when you go to expand your network, you *must* buy the systems from them.

But in an uncharacteristically foresighted mode, the ITU wrote a standard for the interoperability of systems, primarily to overcome the signaling problems. This standard, called H.323, has become the industry standard for VoIP. This does not mean that every VoIP manufacturer builds systems to this standard, nor does it mean that, even if they do, the different systems will be able to talk to each other. But, at the time this book is being written, H.323 is the only standard that has any level of acceptance.

Although other standards, such as SIP and MGCP may have merit, we have made the decision, for the sake of brevity, to only discuss the H.323 standard. The first version of H.323 was intended to facilitate multimedia communications over LANs, but it was subsequently expanded to include support for VoIP over a broader network. This later version is called H.323 Version 2, and popularly referred to as "V2".

H.323 is a very complex recommendation and has incorporated several other recommendations, most notably H.225, RAS Signaling, and H.245 Control Signaling. There are several elements to H.323 that are worth describing, and an understanding of what they do is vital to a basic understanding of VoIP networks.

Terminals

An H.323 terminal is the most basic endpoint for an H.323 network. It can be as simple as a PC that offers voice codec services. Typically, an H.323 terminal is an end user communications device that can encode and decode digitized voice traffic, and send and receive over an IP network. It could include a variety of codecs, including those that support video or other forms of multimedia.

Gateways

Let's look at a gateway that acts as an H.323 endpoint on a local area network. It offers the same services as any terminal, but also serves to connect to other networks by providing real-time two-way communications between other terminals and gateways. It can be the interface between an enterprise network like a LAN and a Wide Area Network, or, more likely, between a VoIP network and a traditional network. These devices can be very small, handling a limited number of connections, or very large and complex, involving multiple networks and on-demand routing.

The conversion needs to be totally transparent. For example, if the gateway is connected to a CEPT E-1 private line, the other end of the E-1 should be completely unaware that the other side of the gateway is part of a VoIP network. Likewise, the VoIP side of the gateway should not have to be at all concerned about the E-1 part of the network. But, terminals and users on both networks should be able to connect at will across the gateway. Most commercial gateways support, in addition to packet switched connections, an array of traditional circuit switched telephony connections so they can easily connect to the PSTN and legacy switches.

The gateway is the key component in a VoIP network, and is the focus of the engineering and thus, the business and sales efforts. It is a critical decision that has to be made early in the planning and deployment of the network. But, unfortunately, it's not an easy decision to make. In general, one would be well advised to go for maximum connectivity to avoid compatibility problems later. One word of caution, if any vendor tries to tell you that their version of H.323 is totally interoperable with any other H.323 gateway, run, do not walk, to the nearest exit!

Gatekeepers

In an H.323 network, the gatekeepers allow a group of endpoints to be controlled from a single point. The gatekeeper functions exactly as the name implies, all access to a network of (say) H.323 terminals, and access within the network is controlled by the gateway. Thus a "zone" is created where the gateway can allow or deny any calls and track all activity. The gatekeeper can also offer address translation services, allowing the terminals within the network to use aliases when communicating with each other.

Those of you who are familiar with TCP/IP networks will recognize that the functions of an H.323 gatekeeper are very similar to those of a router. Within a zone, the gatekeeper can decide who can initiate calls and to whom, when, etc. The gatekeeper can also decide what the outside world can access within the zone.

MCs, MPs and MCUs

An MC is a Multipoint Controller, an endpoint that manages conferences between three or more terminals or gateways. An MC has to manage the dif-

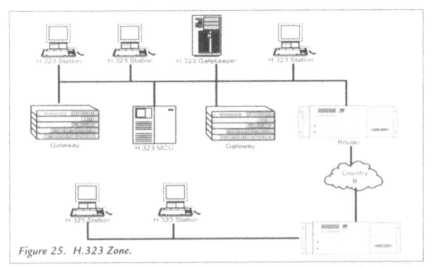

Figure 25. H.323 Zone.

ficulties of monitoring the status of participants and disconnecting inactive parties. An MC is a conceptual entity, and may be part of any terminal or gateway. The purpose of an MC is to control a conference connection, i.e. a connection involving three or more parties.

For each MC, there at least one Multipoint Processor (MP) that operates under the control of the MC and controls the actual data streams. The MP is the switching element of the MC and arranges the interconnection so the parties can all communicate with each other. An MCU (Multipoint Control Unit) is simply an MC and an MP combined together.

Of course, the real power of VoIP is in its ability to communicate outside of its own zone to other zones. This is facilitated by protocols such as H.323, assuming that the gateways can really talk to one another. For instance, H.323 provides several different channels, all within the same communications path, for signaling, control and communication, so only one path is needed.

This is where some of the concepts familiar to those who use the Internet become evident. For example, the same IP addresses that describe endpoints on the Internet are assigned to every entity in an H.323 network. In fact, this IP address may also use Domain Name Service (DNS) so that Uniform Resource Locators (URLs) that form the addressing scheme of the Internet can also be used. For example, a VoIP gatekeeper may have a URL of ras://somewhere@somedomain. This URL, with the standard IP port number of 1719, will allow direct addressing of the gatekeeper. Assuming that the gatekeeper recognizes the calling terminal, communications can be established over any available communications path, including the Internet.

SIP, Softswitches, and MGCP

As you can imagine, in the zany world of telecommunications today, nothing, no matter how good it is, remains unchanged for long. There are just too many things going on in the technology, the business environment and the legal and regulatory scene. Even though H.323 contains all of the technical details and functions needed to thoroughly describe a gateway that would allow voice telephone traffic to be exchanged between it and any other H.232 complaint gateway or switch, the "standard" is lamentably not standard.

Whether it is because H.323 lacks the final definitions and functions that are necessary to make it complete, or whether gateway manufacturers have deliberately thwarted open interconnection, is not relevant. The fact is that H.323 has not become a universally accepted standard, and there is no guaranteed that two H.323 devices made by different manufacturers will actually be able to communicate in a useful way. Amid much finger pointing and gnashing of teeth, it has become painfully evident that the expectations of users is different than what the current crop of devices, taken as a whole, provides to end users.

The world of H.323 is a very frustrating Tower of Babel, where the Holy Grail of interconnection is often tantalizingly close, but still not complete. Devices will connect together, green lights will blink then glow steadily, indicating connection. Pings will be sent, received and acknowledged. All systems will be "GO", but the traffic will sputter and die altogether, with little or no useful indication as to why.

Anyone who has attempted to connect computer networks in the days before a robust and universal TCP/IP, may recognize some of the signs, and cringe, remembering the pain and defeat that lay at the end of many long and hard fought interconnection attempts.

In the early days of telephony, connecting SNA/SDLC networks faced a similar fate. In a effort to trace and eliminate the "trouble" on the non-functioning line, dozens of technicians, test desk operators and engineers often got onto a conference call, and tried to isolate the cause of the problem. As the ill-fated data bits began flowing from one point in the circuit to another, a voice would often declare that the circuit appeared to be working at their location by shouting loudly, "Trouble's fine, leaving here!"

This oxymoronic declaration of self innocence has become the battle cry of the Internet and computer interconnections. But, end users, who don't really care to exonerate anyone when promised connections cannot be made, have sought other solutions when "guaranteed" mechanisms, like H.323 fail.

Softswitches

In the early competitive days, the issue of interconnection of diverse networks first arose, and became either a major stumbling block, or a competitive advantage, depending on how well the emerging carrier dealt with it. Those who were able to arrange the transfer of calls between their budding networks and existing networks had a powerful edge in rolling out new routes. Those who struggled with each and every interconnection attempt seemed doomed to a Sisyphus-like world of constant tweaking.

It was in this environment that the concept of "intelligent networks" first emerged. The idea was deceptively simple, embed enough intelligence into the network that it could dynamically accommodate the introduction of varying transmission types, compression modes and signaling protocols, without the need to individually engineer each and every connection point and virtual route. But the variety of different schemes in the real world made the idea of a softswitch border on being a voodoo technology, and a real world implementation a near impossibility.

Although the range of connection types has not gone down in the last few years, the software and hardware that supports them has become much more functional, bringing the idea of an intelligent network, controlled by a softswitch, into focus once again. There are some real world implementations of softswitches beginning to show in the marketplace. Beware that some of them are vaporware, or provide control over only a limited range of connection types, while promising considerably more. It is a classic case of the marketing hype falling short of the product reality.

While a rigorous definition of a softswitch is still not possible, there are enough successful implementations to suggest what some of the capabilities are. Often called a "gatekeeper", the role of the softswitch is to control call flow over a network, coordinating the activities of a variety of gateways, switches and other telephony network devices. It can be as "simple" as a piece of software running on a conventional PC platform, or as complex as a set of different pieces of software, running on different platforms, either independently or on an existing platform, maybe with an interconnected network of dispersed servers.

When connected all together, the softswitch enables the network to act like a PSTN with Class 4 and Class 5 features, at a minimum. This would include 1+ and 800 service. Again, in the absence of a universally agreed softswitch standard, this seems to be the basis of the standard softswitch feature set, as offered by most vendors. While this minimal functionality does little to advance the state of the art, if it allows this to be applied across circuit switched and packet networks, it has made a significant contribution, nevertheless.

However, if the softswitch is able to play a more significant role in the management of converged circuit and packet switched networks, offering features that are currently not possible, then the contribution of the softswitch technology is major.

SIP

A very popular VoIP remedy to the interconnection dilemma has been Session Initiation Protocol, or SIP. SIP was designed to allow two IP devices to link up over a transmission circuit, and provides high level. It is a very high level, open standard, whose main focus, up to now, has been focused on Internet Telephony applications, like allowing an IP based PBX to communicate with multifunction telephone instruments over a IP circuit.

This means that a PBX would no longer require physical wires or limited range radio waves to connect stations. Any network that supports IP could be used as a transmission path for SIP. A PBX port could be connected at one point in the network, and a SIP telephone at a distant end, even over the Internet, and the SIP telephone becomes a station off of the PBX. Anyone who has tried to implement a remote PBS station would have to marvel at this remarkable ability.

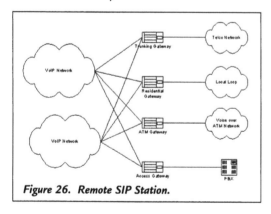

Figure 26. Remote SIP Station.

While this was the intended use for SIP, because of the high level nature of the interface, and the lack of any meaningful limitations on the number or location of connections, SIP came into view as a potential cure for the "Interconnection Blues." As a bonus, many gateway manufacturers included SIP functionality to allow remote PBX stations.

One would have to assume that gateway manufacturers seeking to discourage open interconnections, would be pleased to find out that gateway operators were using SIP to bypass their exclusivity plots. It works, in fact, because SIP is a much tighter and more closely controlled standard, it often works where H.323 does not. Even if the gateway does not support SIP, it is often a simple, and relatively inexpensive matter to add an IP PBX to the gateway, either as an integrated function, or as a standalone device. Although SIP was clearly not designed for the purpose of interconnecting switches using multiple connections, SIP actually works better in many instances than the "standard" H.323 interfaces that were.

Using the Public Internet for Transport

OK, I am going to finally say it. You cannot rely on the Internet for consistent, reliable, high bandwidth transmission of VoIP. There is just too much "bursty" demand for the same Internet bandwidth that we all share. To make the problem even worse, Internet demand is likely to spike at the same time as demand for voice services.

Now, I am going to contradict myself. There are some companies that have implemented VoIP services, which rely on the Internet — with what appears to be acceptable results. Before deciding whether or not to take this approach, let's

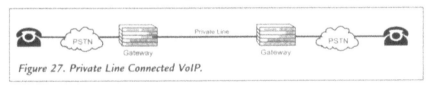

Figure 27. Private Line Connected VoIP.

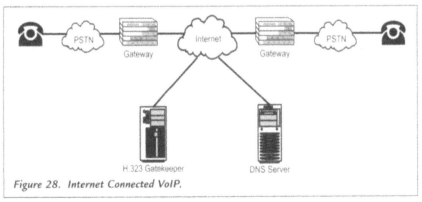

Figure 28. Internet Connected VoIP.

look at two scenarios for transporting voice traffic using the Internet, one that uses the recommended "private lines", and one that uses the public Internet.

Pretty much the same, huh?

The key difference is the guaranteed bandwidth that the Private Line connected VoIP has available to it. Let's make sure that we understand the magic word — THROUGHPUT

Throughput is vastly different from LINE RATE.

There is a misconception that if you want to transmit more data, like if you want to send a digitally encoded voice signal, you increase the line speed. You can do this by getting a faster modem or changing to a different type of line, like ISDN or DSL, which offer greater line speeds. Right?

Well, the fact is that a faster connection device or faster local circuit may not improve the THROUGHPUT at all!

How can this be? If we speed up the connection don't we get a faster overall line speed? Well, maybe, but consider that the rule of thumb is that data can only go over a connection as fast as the slowest link in the communications channel. If you run into congestion at any point, the ISP, the backbone circuit to the Internet, the peering point, or at the far end, the throughput of the channel will be reduced over the entire length of the circuit.

As an analogy, if you take the interstate to work, and it is constantly crowded and moving at 25 miles per hour, will you get to work any faster if you can go down the entrance ramp at 60 mph? Not very much improvement, one would think. You will go the few hundred meters down the ramp to get stuck in the traffic there, and still not reach your destination any faster at all.

Since the Internet is not an engineered service and grew on its own, you do not have guaranteed throughput anywhere along its byways. It is a catch as catch can situation. Sometimes you have very good throughput from one node to another, capable of easily supporting a real time voice channel. Sometimes you have very bad throughput between the same two points, and the delay, or latency, makes the circuit unusable for voice.

You are also sharing the bandwidth of the Internet at every point along the way with anyone else that cares to use it, so you may not get full throughput. A bottleneck along the way will create latency, which may make the circuit unusable.

Chapter 20
Internet Telephony

I have joked many times that the difference between Internet Telephony and IP Telephony is that IP Telephony works! While it may be an exaggeration and oversimplification, if you define Internet Telephony as the making of calls over the Internet itself, then the description is an accurate depiction of the reality. Conversely, IP Telephony uses Internet Protocol, or IP, as the protocol to actually encode and transmit the voice intelligence. The actual transmission medium could be the same private lines as used for traditional telephony, thus theoretically delivering a comparable quality product. Whether it does or not is a matter of debate, and will be covered later.

As discussed, the problem with Internet Telephony is the congestion on "the Internet" itself. This congestion is noted by anyone who regularly uses the Internet for its usual applications, such as browsing the World Wide Web. How many times have you clicked an icon, only to have to wait a considerable length of time before you get the intended response? Not all of this is due to the slowness of the server or the application on the other end. Much is due to the latency in the Internet itself, which gets worse as traffic builds.

Aggravating the situation is the increasing use of high bandwidth content on the Internet, like multimedia, audio and video information. In fact, the use of the Internet for voice telephone traffic will, itself, tend to lower the quality of the calls due to the impact of the increased demand for bandwidth caused by such telephone-related traffic.

It is because of this lack of consistency that I refer to Internet Telephony as "ham radio". If you have ever had a discussion with a ham operator, they often talk of clear conversations to distant locations around the world. "Can you talk to Moscow?" "Yes." "Right now?" "Well, no, we have to wait for the sunspots to peak and for the MUF of the F layer to reach 14 MHz, then we send out some CQs if there is anyone there that can hear us. Of course, if the QRM is too great, we will not be able to converse, even if the signals can make it."

You hear the same sort of explanation as to why Internet Telephony sometimes works and sometimes does not. On the other hand, if you use some guaranteed bandwidth transmission scheme, like private lines, you *can* use IP telephony to make a solid connection every time.

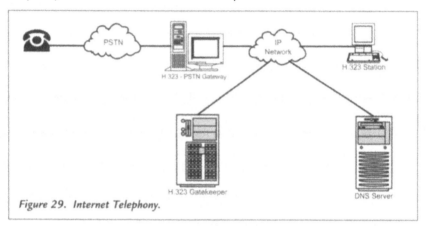

Figure 29. Internet Telephony.

Still, someday — maybe — you will be able to actually use the Internet to complete telephone calls end to end. More and more bandwidth is coming on line all the time, making it more feasible. But, with pure Internet Telephony offerings, there's a unique problem — what happens if the other end of the connection is not on line at that moment?

Ever use an Instant Messaging service, like AOL? You have a "buddy" list of the people you want to talk to, and IF they are on line, you can call them at that moment. If not, you can't even leave a message. And, if the person you want does not subscribe to the service, you cannot contact them at all. Many of these same constraints apply to the pure Internet Telephony services.

So, it is obvious that there will have to be some connection to the PSTN to make Internet Telephony a viable offering. This can be done, but the charge for

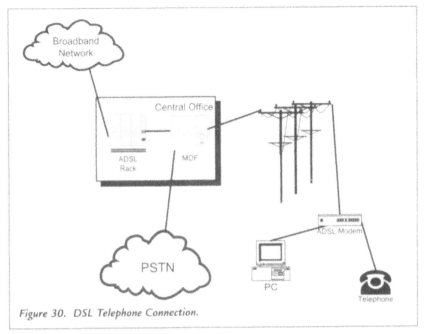

Figure 30. DSL Telephone Connection.

getting from the IT node to the end user may be just as expensive as the cost for just calling them directly. The real kicker in all this is that the cost of long distance is dropping so fast (in the US the 5 cent minute is common) that the motivation is not as strong as it was a few years ago, when long distance rates were much higher.

What may happen, however, is a melding of Internet access and local telephone service into a common facility. This is already happening to some extent with DSL, where the data is physically multiplexed over the same pair of wires

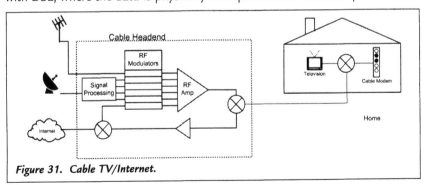

Figure 31. Cable TV/Internet.

that the local telephone service uses.

Cable offerings, like the immensely popular Road Runner Internet access offering, have already combined cable and Internet. Again, this is a form of physical multiplexing, since the Internet data simply uses a cable TV channel to send and receive Internet data. There is no logical connection between the two.

It is conceivable that in the not too distant future, ALL information services may be sent over the same communications path. This has many advantages, but there is also one notable disadvantage.

The advantages are pretty significant. Since TV, telephone and Internet are riding on the same communications path, wherever you have an information outlet, you could have a device or devices connected, be it a telephone, a television or some data device.

Having all these services together allows other interesting possibilities as well, like ordering pay per view over the data device. Or, if it is one of the computer TV combinations, you could order and watch the event on the same device. There are lots of intriguing possibilities here, and lots of potential for making money.

The main drawback is that in the event of a disruption in the facility, you lose everything. Even today that can be a problem. Bellsouth, for example, has an online trouble reporting service on the Internet. But, if you are using Bellsouth for your local service and DSL, how do you report an outage? You cannot call them on the telephone, because it is down, and likewise, you cannot report the problem over the Internet because that is down as well.

(In the case of Bellsouth, they have even made it impossible to contact their repair service over a cellular telephone, so you have to go to a neighbor's to report the outage!)

Chapter 21
An Introduction To
Voodoo Technology

This is a new industry. As such, it is full of promise, full of opportunities, but also, as with anything new, it is full of unforeseen pitfalls, drawbacks and unknowns. The legions of snake oil salesmen, who specialize in exploiting the unknown and unknowable, abound. Whether they deliberately lie to sell their product, or lie because they don't understand the issues, or whether others have mislead them is irrelevant and a judgment call that doesn't need to be made. It's the impact of these technological lies that's relevant.

The best protection against them is a solid base of knowledge on which to make judgments; and a "show me" attitude when faced with "incredible, breakthrough technologies". These two things should provide a "technological condom" keeping you safe, secure and uninfected by broken promises.

Recognizing When to Question

It is very difficult to distinguish between what is real and what is marketing hype, a false promise and/or an outright lie. We are all allured by the promise of new technologies, and in fact, some of the new technologies appear, rightfully so, to be nearly magical in their functionality. We want to believe; we need to believe. This desire makes a fertile ground for those with less than honorable intentions to plant their seeds of deception.

The original high tech salesmeister, Thomas J. Watson, once said that any salesman who did not tell at least two lies every day was not doing his job. This greatly misunderstood quote was probably taken out of context and he probably did not mean to instruct his followers and disciples to deliberately say something that they knew to be false. What he meant to say was that sales representative should be very aggressive and positive, and not let a gap in the salesperson's knowledge impede the speedy closing of the sale.

The damage was done, however. Once the words were uttered, the fast paced high tech sales culture was off and running. The world has undergone a revision of its moral underpinnings in the interim — at one time, taking an educated guess, or applying the best possible spin to a situation, was considered lying — today it is the accepted norm. It is far more important to make the sale quickly.

Against this backdrop of "truth slip" in morality, our industry struggles with a plethora of bad information and Voodoo Technology. Before you run off to get your dictionary, Voodoo Technology is not yet there, so I will undertake the first formal definition of this term.

> **Voodoo Technology** *n* **1.** A pseudo science encompassing concepts that are compelling and believable, but currently beyond reality. **2.** The offering of products or services that are widely perceived as necessary and desirable, but without the ability to deliver these products or services. **3.** Use of techno·babble to deliver general confusion about the scientific basis for incredible claims of functionality —**Voo·doo Tech·nol·o·gist** *n*

This term will undoubtedly be given additional dimension and meaning as our industry unfolds, but there is already an impressive list of Voodoo Technologies that are out there today. Some of these are:

Voice Compression Equipment: Equipment capable of 64:1 or even higher ratios, with no loss in quality. This remarkable technology is attractive. Because, if, make that, IF, it could be shown to actually work, then a single 64k circuit could easily accommodate 2 E-1's, or over 2-1/2 T-1s. Think of the possibilities here, a 64k circuit, even a dedicated line is very cheap when compared to E-1 or T-1 circuit costs. It is usually 10 percent, or less of the cost of a T-1.

So, if this worked, we could send 64 voice channels over a single channel, called a DS0!

Well, the truth is, you could compress 64:1, or 128:1 or 10,000:1 if you wish. The problem is, something has to give. Compression is simply reducing the number of bits used to represent a voice signal, and always involves a trade-off between lowered bandwidth and quality.

Try it for yourself. Take one of the familiar MP3 "rippers", like MusicMatch, and try ripping a CD at different compression rates. Take off a song at 160 Kbps. Listen to it. Now try 128 Kbps, how does it sound? Probably not a lot different, maybe not quite as clear, or maybe some of the highs are lower or a bit distorted.

Now try 96 Kbps. Even a casual listener can hear the noticeable drop in quality. The file will be smaller, by 40% or so, but you can definitely hear what had to give for this savings. Now try 48 Kbps. The music begins to sound like a cheap transistor radio, i.e. much distortion — you can still hear the music, but you would not want to listen to it for very long. However, the file is now a whopping 80% smaller.

If you really want to see the effects of compression, now go to 5 Kbps, and the music is recognizable, but not really useful for anything. The file will be over 90% smaller than at 160 Kbps, but the price that was paid in quality probably makes the resultant file unusable.

In this demonstration, we went from an encoding of 160 Kbps to 5 Kbps. This is a change in compression of only 25:1, and you can readily see what the effect on quality was. Can you imagine what the result would have been if we had gone to 64:1? Remember, with voice communications circuits, we are not starting with a 160 Kbps channel — more like 64 or 56 Kbps, so the drop in quality will be even faster.

The lesson here is that you have to actually try a technology as critical as compression to see what it sounds like in actual use. Actual use does not mean a booth at a trade show, either. You must see it in use in its intended environment — connected to PSTN circuits, etc.

Satellite Echo Suppressors: Remarkable devices that not only completely

Format	PCM	mu law (G.711)	MPEG-1	ADPCM (IMA/DVI)	GSM (6.10)
File extension	.wav or .aiff	.au	.mpa or .mp2	.wav	.gsm
Data rate	128Kbps	64Kbps	32Kbps	32Kbps	13.2Kbps
Bits per Minute	960K	480K	240K	240K	96K
Compression factor	1:01	2:01	4:01	4:01	10:01

Figure 32. Compression Ratios vs. File Sizes.

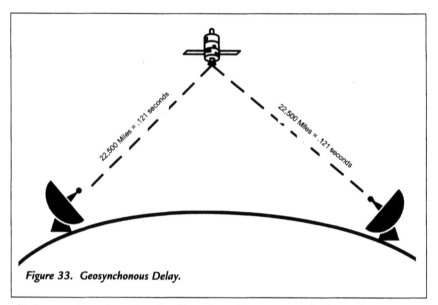

Figure 33. Geosynchonous Delay.

eliminate echoes, but also significantly reduce or even eliminate the propagation delay on long haul routes. Why not? It seems perfectly reasonable that we
could do something to lessen the delay, right?

No, you cannot reduce propagation delay, whether it is satellite, digital
radio, or even fiber. The problem is most acute on a geosynchronous satellite
circuit that's going to travel 45,000 miles or more; it will take a fraction of a
second to make the trip, no matter what. International carriers have almost
completely moved their voice traffic off satellites for this reason.

How are you going to reduce the time delay? Look at Figure 36. Logically,
there are only a few things you can do to reduce this delay. First, you could
speed up the signal. Well, as far as anyone knows, the speed of light is a constant. Einstein would roll over in his grave if he thought you were monkeying
with that.

Secondly, you could start sending the signal *before* the speaker actually
spoke the words. Well, short of an Amazing Kreskin trick, I cannot imagine
how you would actually accomplish this. Finally, you could lower the satellites
to reduce the distance the signal must travel. This will actually work, and is
why Low Earth Orbit (LEO) satellites are used for voice services by ICQ,
Iridium and Globalstar, but it took a constellation of over 60 satellites to do
it. (All three went out of business, also.)

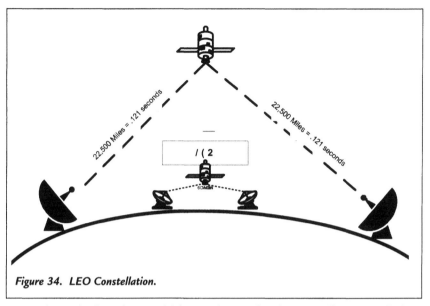

Figure 34. LEO Constellation.

To show the lengths to which Voodoo Technologists will go to sell their products:

A few years ago, I was invited to a demonstration in Geneva, Switzerland by one of the top 10 communications equipment companies, who claimed to have a technology that would allow them to reduce propagation delay.

The demonstration was remarkable.

They had set up a simulated satellite circuit that introduced a 1/2 second delay each way into a communications circuit. There were two telephone booths, and they had an editor for a major communications magazine go into one booth, and I went into the other. The booths were arranged so that we could not see or hear each other. The presenter let us begin talking over the circuit, with the delay. It was, as expected, very difficult to hold a coherent conversation without resorting to saying "over" when you were done talking.

Then they switched in their incredible new equipment, and the delay was gone. My friend and I were able to converse without a noticeable delay in the signal. The short demonstration was numbing.

We were amazed, and suspicious. It defied logic. So, we hid out in the convention hall, until after it closed and everyone was gone, and snuck back to the booth to see how this was done. We entered the telephone booths, and began talking. But this time, we did two things differently from the demonstration that afternoon. First, we left the booth doors open so we could see, and hear, each other. Secondly, we talked on for a while.

During the afternoon demonstration, the presenter had led us through the conversion, not allowing us to speak for long, interrupting one to have the other begin speaking.

The trickery became obvious. What they had done was to introduce a bit of delay in the "white space" between words as the voice was reproduced on the opposite end. This had the effect of stretching out the conversation. In fact, if one end spoke for more than 20 seconds or so, the effect was clearly heard. Words began to have several seconds of silence between them, to allow for the delay. The demonstration in the afternoon had been carefully managed to make sure that neither of us spoke for more that a few seconds, minimizing the impact of the delay.

Clearly, this "technology" would not be usable for anything beyond a hokey trade show trick, but they were attempting to actually sell this technology. The next day, my friend and I again volunteered to demonstrate the device, but resisted the presenter's attempts at curtailing our conversation. The audience started laughing as the voice on the speaker began to roll out like a bit from Bob and Ray's "The World's Slowest Talking Man."

Fortunately, this sham was exposed, and was never seen again, but there are many more where this came from, so beware!

Fax Compression Equipment: Technology that can compress already compressed Group III faxes to even greater ratios, allowing 8:1 or more compression of faxes.

This one is not quite as evident as with voice: it is far more difficult to compress something that is already compressed, like a Group III fax. All the obvious things, such as taking out blank lines, blank spaces and repeated sequences have already been done, and very efficiently, in Group III.

TINSTAAFL!

There Is No Such Thing As A Free Lunch! And faxes cannot be compressed. Faxes are one of the most troublesome things for service providers to accommodate. They are frequent, random, and worse yet, highly sensitive to attempts at compression. The fax machines will doggedly try to stay with an inferior circuit, testing the circuit and adjusting at every page, disconnecting only when all else fails.

OK, but why won't a fax compress like any other analog signal?

Simple, it is already compressed, the white space removed, some pattern recognition done and most of the other bandwidth wasting items fixed. It will not compress any more, and any attempt at doing so will cause it to slow down, extending the time it takes to transmit. If it cannot get at least 4800 bps, it will disconnect. And, it will measure the quality of the circuit at every page.

What this means is that a different strategy needs to be developed to handle faxes. One common strategy is to have the gateway on the originating end "pre-indicate" the fax, which is simple to do. Fax machines send a CNG (for CalliNG — we did not make this up!) tone when they call out, so that the machine on the other end will know that a fax machine is calling and will answer. It is also what allows an "all-in-one" telephone device to determine if the incoming call is a fax, or a voice call.

It can also cue the carrier that a fax is being transmitted, and that additional bandwidth will be needed. The disadvantage of this method is that a portion of the bandwidth that the carrier has must be reserved for this one call; although the fax can proceed unrestricted.

A cool tip: if you think your carrier is using this fax detection scheme, generate a 1,100 hertz tone every three seconds while you are placing the call, and you may get the best bandwidth the carrier has to offer. If you have perfect pitch, 1100 hertz is D flat, or MIDI tone 85.

ANI Recognition Software: Technology that can detect the calling party's telephone number, even when the originating telco does not bother to send it. The ANI recognition rate is vastly improved over what it was a few years ago, and gets better all the time, but the fact is, ANI is not even sent reliably on US domestic calls, much less on international calls. In addition, there will always be some locations that will block, omit or deny access to the ANI information, so if the communications scheme relies on ANI, it may be in deep trouble.

There are very few countries that have agreed to consistently provide ANI information outside their own borders. If the originating gateway does not send the ANI, nothing short of ESP can retrieve it.

You may look at this list and laugh at some of the items, or you may wonder why other people are laughing at other items, or you may even get defensive, if you think you have or want any of these items. But be assured, that if someone promises you something on this list, begin asking a lot of questions, because you may be talking with Thomas J. Watson himself.

Free Bandwidth: Bandwidth that will allow the carriage of unlimited telephone traffic over the Internet. Again, the principle of TINSTASFL tends to make this impossible. If transporting wide bandwidth data over the Internet were possible, why would carriers spend time and money securing bandwidth? The truth is that you cannot rely on the "free" Internet bandwidth as a primary strategy. That's not to say that you can't at times use the Internet for this purpose.

Remember a few years ago when the Monica Lewinsky papers were put onto the Internet, and people around the world, eager to learn all about this scandal, accessed those voluminous files, and clogged the Internet for days? Well, if you were using the public Internet to transport your traffic from Bangladesh to the UK, you would have been out of business until everyone's curiosity was satisfied.

This does not mean that the "free" bandwidth on the Internet should be shunned. On the contrary, it *can* be used, but it has to be carefully planned and managed. The first step is to devise a rigorous *on-going* testing program. By using existing Internet TCP/IP tools, it is possible to accurately measure the quality of the route to the other end. Since this route will dynamically change, a trending should be established. This will reveal the overall quality of the route, and problem times and days. Once the characteristics are completely understood, it is possible to utilize some of the bandwidth.

This may help with an understanding of how private backbones and guaranteed bandwidth works:

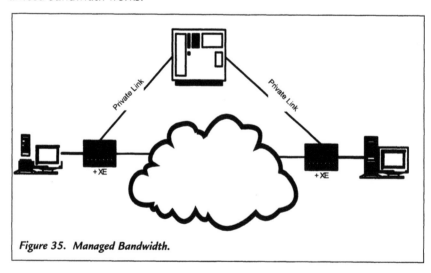

Figure 35. Managed Bandwidth.

One of the very powerful concepts in the Internet's TCP/IP structure is the capability for routing packets in alternative ways. Keeping traffic off the public Internet routes, and transporting over private backbones is possible, and can offer a mid-level alternative between private lines and the public Internet. Managed bandwidth suppliers can offer functionality that will allow the use of quasi-public facilities to carry critical traffic.

The H.323 Standard: A standard that allows any terminal or gateway to talk to any other. One of the great mysteries of the universe is standards, that aren't standards. H.323 is one of those. Supposedly, if you use H.323, any gateway can fully communicate with any other gateway. It turns out that this is not true because there are so many exceptions to the rule. Regardless of the idea of compatibility, the reality is that each gateway has to be certified to interoperate with any other.

What H.323 will do is allow two disparate devices to recognize each other and to establish and manage a communications path between each other. What it does not do is to guarantee that this path will be useable for carrying telephone traffic. There are also many other issues that must be addressed, like signaling, compression and control.

Why, you might ask, would manufacturers not want their boxes to be able to talk to everyone else's box? This is a very good question, and one that transcends telephony. Every high tech manufacturer embraces standards. In fact, they *insist* on them. But, they want their proprietary scheme to be THE standard, if they are a market leader. If they are not market leaders, they want an INDUSTRY standard that no one company controls.

This is the eternal struggle that preoccupies the minds of corporate executives and technological innovators. Every manufacturer wants to lock in its current and future customers and to be the one that everyone has to build to. It is usually only when chaos threatens that they get together and work out a standard. There are ample examples of this outside the telecom industry. Consider the video tape wars of the mid-80s — the struggle between Sony's Beta format, and the VHS standard waged on for years. In the end, Sony lost the battle, but not the war; it just turned its considerable manufacturing clout to VHS. Had its Beta format won, everyone in the industry would have been forced to concede development of videotape to Sony, giving it a formidable market advantage.

It's the same in the establishment of VoIP standards. H.323 is just one step. It does not guarantee interoperability; although it is the first step towards standards that may someday allow the exchange of VoIP traffic to be as predictable as it is with legacy systems. That's difficult.

Section IV

Legal and Regulatory Issues

Chapter 22

A Regulatory Perspective

Things have changed radically over the last few years, but telecom's global environment today is anything but completely liberalized. It is true that many of the legal barriers have been lowered, but in many cases, they have been replaced by practical considerations that are sometimes even more restrictive. In the Caribbean for instance, the UK company, Cable and Wireless, still has such a death grip on telecommunications, that it is still impossible even in this new Millennium to find an alternative carrier. Many of England's first New World colonies were in the Caribbean, and it appears that this English company has no intention of letting the sun set on its corporate flag in the region anytime soon.

In Europe, despite a European Union mandated opening of telecom markets in 1998, there are still *de facto* barriers to entry — many so onerous that they bar real entry to all save a chosen few. In Germany, alternative carriers must install so many switches to get a license, that only 3 or 4 carriers in the world have the resources to meet the minimum requirement.

In Asia, Japan is still struggling to protect its incumbent carriers, and in China, they are searching for answers on financing the expansion of the world's largest potential market to the world's most underserved populace, and still not give it meaningful competition.

In the face of the considerable issues involved, and the recognition that telecommunications is as basic a "natural resource" as any other in today's world, policy makers are radically changing the way regulation is approached. Telecom regulation is a needed tradition, given the public nature of the service, but telecom cannot thrive without the stimulus of market competition. The experience in the US should be adequate proof to all but the most cynical monopolist.

It is admittedly a juggling feat of great proportions to balance and manage the metamorphosis of traditionally monopolistic utility type services into a modern, high tech, fast-paced competitive enterprise. The regulators and policy makers are key to the success, or failure, of this effort. Mistakes made are not readily apparent, and may shadow the industry for years to come. The temptations and pressures that are brought to bear by influential and powerful interests (often the very government itself) to make short term, immediate profit-oriented decisions, or non-decisions, is enormous. The ability to see the future, and commit to it is a rare trait, one that appears infrequently, if at all.

As with any political process, this crucial procedure is infected with special interests that can often win by simply delaying a decision. AT&T did this for years in the US, and made a lot of money along the way.

There is no one way to plan the deregulation and liberalization of a previously monopolistic industry, and there are many outside influences that will impact any such plan anyway. Techniques like callback and VoIP obviously forced the decision in many locales, where the tendency would have been to just sit back and wait. Both of these technologies introduced competition that would have otherwise been put off for years, or even decades.

Both callback and VoIP pushed the envelope by using the existing infrastructure to provide competitive service where they were not welcomed. They share a stealthy nature, and are difficult to detect and disrupt, even when the legal environment allows such interruption. Callback uses existing, regulated structures in a clever way. The only complaint, other than a frank admission that competition is unwanted, is the unanswered trigger call. This tenuous argument only resonated where there was a predisposition toward acceptance of such a flimsy excuse.

VoIP uses the generally accepted principle that data isn't as regulated as voice, so its "data" makeup allows some latitude to be granted in the name of the advancement of science. In the EU, Internet services were officially excluded from the regulatory regime thus allowing VoIP to sneak in "under the radar."

In the early days of callback (after the AT&T led effort for an outright ban failed), those opposed to competition made the decision that enforcement of competitive telecom bans in their own jurisdictions would be troublesome at best or cause a backlash, in a worst case scenario. So, these regulators, reluctant to start a war in their own jurisdictions, often took the battle outside their countries, mounting their attack on competition at the FCC. Their theory was that it was the "rogue" US carriers that were causing the problem in the first place, so why not attack the problem at its source. Unfortunately (or fortunately), the US Federal Communications Commission, on a crusade of its own to export the US brand of telecom competition, gave only lip service to enforcing the questionable regulations that were designed to thwart competition.

In fact, the FCC formalized their procedures for handling these complaints in a decision that put the onus back on the country that wanted US action. The decision required that the foreign regulators demonstrate that they had made a good faith effort to enforce their own regulations in their countries. Only if this effort failed because of the actions by US callback companies, would the FCC even consider taking any action. In all the years, and numerous petitions, only one formal action was ever taken, half-heartedly, against a US callback company. Even then, it is more likely that the FCC was motivated by a different agenda and gave vent to a desire to punish a couple of specific carriers for things unrelated to callback.

Chapter 23
Privatization vs. Liberalization

Telephone competition in the United States brought arguably better service and unquestionably lower rates to the long distance market. There is little doubt that a similar opening in the telecommunications market is coming now to the worldwide scene, regardless of how much powerful interests may wish that it would go away.

It is too compelling an idea and offers too many economic benefits to a country to be held off for very long. The question is, how, and under what conditions will individual countries open their telecommunications markets to competition.

If you want to be a player in this global arena, you must understand how competition is going to unfold in order to design strategies, market plans and technical plans to position your products for this evolving market place. And the market is huge. By the mid-1990s, the international long distance market was over $60 billion annually, and growing in excess of 10% annually. Today that market is closing in on the $90 billion mark. Entrepreneurs, investors, consumers, regulators and telephone companies all view this trend from different perspectives and with different interests.

Investors and entrepreneurs dream of new global mega-corporations being spawned in the primordial soup of a competitive international telecommunications marketplace, much as MCI, Sprint and others began their rise in the early seventies when the US market opened.

One of the most misunderstood forces affecting open competition is the privatization of the PTT, often seen as a prelude to opening a market to competition.

Foreign regulators, who work for the same employer (the national government) as the company being privatized, i.e. the PTT, worry about what the government REALLY wants them to do. Although the telephone companies (i.e. PTTs) see the growth in telecommunications that competition brought to the United States, they have a knee-jerk adverse reaction to the concept of open competition. In their mind, there is a risk that they could lose.

FCC Officially Fosters Competitive Environment

It is the official policy of the Federal Communications Commission and the US government to foster open markets and competition around the world. In the area of telecommunications, the United States has an even greater interest, because of its advantageous position on the supply side of a competitive equation, thereby enhancing its own balance of trade. When the FCC and/or the United States government prod a foreign country to open its telecommunications markets, the foreign response is often that it is in the process of privatizing its PTT.

Is this a valid response to the demand for open markets, or an artful dodge? The argument from the PTT, and thus the foreign government, is that the two concepts are linked — privatization is necessarily for competition to begin. Does it not seem logical, on the surface, that to allow competition, we must first privatize the PTT? Otherwise, new entrants to the marketplace will find themselves competing with the government, which also controls the regulators. But is privatization of the PTT really a necessary prelude to the opening of the global telecommunications market?

Let's first look at what happened in the United States when it opened its markets. Keep in mind that the United States had a unique structure with its own PTT — AT&T was a privately owned telephone company. The government controlled the regulatory agency (the FCC), so its decisions were independent of the ownership of the PTT, and oriented to the needs of the people, not tied to its revenue structure, as it is in the rest of the world.

Would the rollout of the US domestic competitive marketplace have taken a different path if the AT&T stockholders had controlled the FCC in the 1970's? You can bet it would! AT&T, and especially its stockholders, fiercely fought the rise of competition. They acquiesced only when faced with what appeared to be a no-win situation. They decided to negotiate to see what they could salvage before decisions were made without them. Even afterwards, AT&T continued to fight the regulatory underpinnings of competition whenever the opportunity arose.

PTT Reaction and Response

Is there any reason to believe that a newly privatized PTT today will react any differently than AT&T did back then? Probably not, since they are just as concerned about their revenue stream as AT&T was. Consider also, that in most instances, the government does not really sell all of the PTT, nor does it relinquish any significant control in it. In most countries, privatization really amounts to the government selling bonds in the PTT. It amounts to a fund-raising venture, with the government retaining the majority ownership of the PTT.

In fact, the government often feels a need to protect the newly privatized PTT, so it usually takes a more anti-competitive stance than would have been deemed to be reasonable otherwise. In Singapore, for example, the PTT, Singapore Telecom, was given a legal monopoly for three years after privatization was completed in 1996. Other countries have taken a similar tact, some giving the privatized PTT up to seven years. The official reason for the delay is to get the infrastructure in place to allow the PTT to effectively compete. The real reason is probably to give the PTT a head start on competition, knowing that it will be at a disadvantage when it does begin to compete in a more open marketplace.

In Ecuador, announcement was made concerning the move towards privatization of the PTT, and even before any reference had been made to competition, the union threatened a strike if competition was allowed. The union's threatened action, which occurred during the 1994 presidential elections, not only derailed the privatization process, but elevated it to a national political issue. The PTT's upper management was replaced by a structure thought to be more accommodating to the union's point of view. Privatization has not made much progress since then, and is still being "studied." Further delays will likely ensue.

The main point to keep in mind, as the world telecommunications market unfolds, is to not be misled into equating privatization of a PTT with open competition. They are entirely separate concepts, not at all related to each other, except that the government uses privatization as an excuse to preclude competition. The government prefers to deal with privatization because it can completely control the process. Competition, once it starts, will be under the control of market forces and the government's ability to control it will be very limited.

With an understanding of these concepts, you are in a much better position to understand the events as they occur. You have some important clues as to which countries to target and how to serve them.

If the PTT is privatized, or in the process of privatization, look for interim anti-competitive measures, including the blocking that is happening now in much of Central America. In Central America, where privatization is in the early stages of rollout, the perceived need to protect the PTT is resulting in many technological attempts to thwart resellers. If privatization has not yet progressed beyond the planning stage, there may be a more lax environment in which to operate. Older rules will tend to be in effect, and the market may be a bit more open.

Once a country approaches the end of the privatization cycle, things might open up a little more, at least in theory. As an example, in England, where privatization is nearly complete, and a second telephone company, Mercury Communications, is in operation, you will find some of the most "closed" public and pay telephones in the world. You cannot normally make a toll-free call from either a British Telecom or a Mercury pay telephone. They disconnect toll free numbers after three touchtone digits are heard. Hotels routinely block all toll free calls, precluding unencumbered access to airlines, the national rail system and even to BT itself! Operator intervention is required, throwing toll free telephone technology back 20 years.

Designing an Effective Strategy

It may pay big dividends for any potential international long distance reseller to be aware of the forces shaping the rollout of competitive telephony in any country, and the status of privatization in the target markets. Knowing exactly where the country stands in the process of privatizing its PTT markets will give a strong indication of what the nature of the response to a competitive threat may be. This will allow the service provider to tailor the offering, design the access method and determine the pricing points. Just as a service provider must analyze the PTT rates and the state of the national telephone network in deciding how to roll out services in a given country, they must also analyze the legal and regulatory environment. Knowing the status of privatization will provide many clues to the challenges and opportunities awaiting in that country.

Chapter 24

Accounting Rates — A Major Issue

The major financial issue facing the future of international resale of long distance services is the accounting rate. The so-called accounting rate is the way that the foreign PTT and the US carriers agree to split the revenue for calls that originate in one country and are completed in another. An understanding of how the accounting rate works and its impact on different parties is vital to setting a strategy for callback, whether you are providing it, buying it, battling against it, or whatever your involvement. If you are planning for the future, you must understand the impact of accounting rates or you simply won't be able to predict the trends.

The issue of accounting rates is a very confusing, complex and contradictory one. I apologize in advance for any head scratching this chapter may cause you. It was simpler when each country had only one carrier. In fact, in some ways accounting rates still operate as if there were only one carrier, even though that's not the case.

Let's go back in time for a moment, to when the world of international long distance was simpler — you had one agency in each country that had the authority to set rates for long distance calls, both domestically and internationally. This agency was free to negotiate whatever terms, conditions and rates it deemed necessary.

In those days, the PTTs simply agreed to a fixed, usually per minute rate for calls between countries, regardless of the in country rates for long distance. This is called the "collection rate". Also, in those days AT&T did the negotiations for the US, since it was the only long distance carrier, thus making it the US PTT.

The two PTTs would take a look at the amount of traffic flowing between the countries, the costs for providing the facilities and switching the calls. They would then add an appropriate (or totally, inappropriate) margin of profit and set a per minute rate, based on some currency, for calls between the countries. This is the *Accounting Rate*. Each country is entitled to one half of this amount of money. This is what is referred to as the *Settlement Rate*, which is equal to one half of the accounting rate.

In theory, this rate takes the call to a point halfway between the countries. This could be in the middle of the ocean, but it is only a conceptual point, so there is no equipment associated with it. It does not matter in which country the call originates.

The PTTs then keep a dual column log sheet of the calls and the resultant revenue. At the end of a set period, usually quarterly, the calls are reconciled, and the two PTTs split the revenues. If the number of minutes from each country is exactly equal (which they never are), neither PTT would owe anything. But, in reality one PTT always ends up paying the other a settlement. The PTT with the greater amount of originating traffic pays the other PTT an amount equal to the settlement rate times the number of excess minutes.

As an example, let's use the following scenario. In a given month, Parador sends one hundred thousand minutes of traffic to Transylvania, and Transylvania sends one hundred ten thousand minutes of traffic to Parador. Thus Transylvania has sent an excess of ten thousand minutes to Parador. The accounting rate is one dollar, meaning that the settlement rate is fifty cents. Transylvania, then, for this month owes Parador fifty cents per minute for ten thousand minutes of excess traffic, or five thousand dollars.

The other way of looking at this is that Parador has generated $100,000 worth of traffic at the set accounting rate, and Transylvania has generated $110,000. The total is $210,000, to be split between the countries, or $105,000 each. Parador collected $100,000 of this mutual pot and Transylvania collected $110,000. Thus Parador is short $5,000 and Transylvania has an excess of $5,000, so Transylvania sends Parador $5,000, and they are even.

In this example, the PTT in Parador has agreed to a $1.00 Accounting Rate.

Parador		Transylvania	
Minutes to Transylvania	100,00	Minutes to Parador	110,00
Excess Minutes	0	Excess Minutes	10,000

In this example, the PTT's in Parador have agreed to a US $1.00 Accounting Rate.

Transylvania sends 110,000 minutes of traffic to Parador and Parador returns 100,00 minutes of traffic. This gives Transylvania a 10,000 minute overage of traffic.

Transylvania then has an excess of traffic, or debit of 10,000 minutes times US $.50 (settlement rate =1/2 the accounting rate) or US $ 5,000.

Transylvania thus sends US $ 5,000 to Parador to balance the books.

Figure 36. Accounting Rate Example.

Transylvania sends 110,000 minutes of traffic to Parador and Parador returns 100,000 minutes of traffic. This gives Transylvania a 10,000 minute overage of traffic.

They then balance the payments by crediting 100,000 minutes times $1.00, or $100,000 to each party. Transylvania then has an excess of traffic, or debit of 5,000 minutes times $1.00 (settlement rate) or $5,000. Transylvania then owes $5,000 to Parador to balance the books.

In the meantime, however, each PTT collects from their own customers, but not necessarily at the same rate. The amount that they collect from their subscribers is called the Collection Rate. This collection rate is not part of the formula. Both PTTs are free to set the collection rates at whatever level they wish. What is important to note is that, in essence, each PTT is paid the settlement rate to complete the calls to its own customers.

When each country had one carrier, it was a simple issue, and the settlement process was easy. With the advent of multiple carriers, the issue became much more complicated, and even affected the way in which calls were routed and carried between countries. But, in essence, the same settlement process would occur between every carrier in every country with every other carrier in every other country. There are approximately 240 nations with PTTs (or a PTT equivalent) in the world. If there is one carrier in each nation, it has to settle with 239 other PTTs to complete the process.

One important point before we move on to proportionate return issues is that since the advent of competition in the US, each carrier, at least in theory, makes its own agreement with every other carrier. But one of the rules of engagement is that foreign carriers are not supposed to discriminate against carriers in another country. So, if AT&T makes an agreement with the PTT in Parador, that same accounting rate should apply also to Sprint, MCI Worldcom, etc. But because of time delays in completing agreements it tends to be a dynamic process. At a given point in time, the rates may not, and probably won't be, the same for all services.

It is also important to note that different services can, and usually do, have different accounting rates. IDDD (International Direct Distance Dialing) is usually far less than 1-800, or International Toll Free Access, even though they use the same facilities. This is one tool that the PTTs use to prevent the spread of international resale of long distance. Raising the accounting rate of toll free access eliminates a potential source of competition by increasing the cost of using toll free numbers for access. In many cases, the accounting rate for toll free access is 2-3 times the accounting rate for IDDD.

The impact of the accounting rate on competitive services is drastic but the disparity in collection rates is what fuels the fires of entrepreneurs and makes it profitable; both to end users and service providers.

✎ Chapter 25
Proportion Return

Now that we have established how the carriers in different countries divide revenues, let's move on to the important topic of how they divide the traffic when multiple carriers are involved in one country or the other. This may be an issue that is not familiar to non-US consumers, since they lack the ability to select their own carrier. In fact, with equal access and 10XXX dialing in the US, we can, and often do, pick a different carrier for each call. The Least Cost Routing feature available in PBX's sold in the US since the early seventies is useless in most foreign countries. There is only one route for each and every call, local, domestic or international — the PTT.

So every call leaving Parador goes through the Parador PTT. Easy enough, but what happens when the PTT in Parador has a choice of carriers to send a call to in a given country? Since, at least in theory, every carrier is on an equal footing and has equal rates, there is no price advantage for using any one carrier, so how is the carrier selected in the other country, if there is more than one?

By convention, traffic out of a single carrier country into a multiple carrier country is divided on the basis of "proportionate return". It is a simple principle. The PTT in the monopoly country divides its outbound traffic in the same proportion that it receives the inbound traffic. So, if Carrier "A" is responsible for 30% of the traffic coming into Parador from Transylvania (where there are multiple carriers), Carrier "A" will be rewarded by receiving 30% of the traffic coming out of Parador into Transylvania.

In the case of a US-based carrier into a non-competitive country, this can be a windfall. If the accounting rate is $1.00, the US-based carrier will get $.50 a minute for foreign traffic coming into the US. Since the cost to switch and transport calls in the US are well under a dime a minute, there is tremendous profit in these calls.

This is why you will sometimes see US carrier rates to a particular country that are less than the accounting rate to that country. Even if the carrier loses $.05 a minute on traffic going into a given country, the carrier could net $.45 a minute on the calls out of that country into the US. This is why you will rarely see a US-based carrier complain openly about high accounting rates — they benefit from them as much as the PTT does. As we will see in the next chapter, there are US government concerns about high accounting rates.

It is also important to note that IDDD traffic is the only type of long distance that counts toward the proportionate return and does not normally include 1-800 traffic. As mentioned, 1-800 traffic might even be disallowed or discouraged. Although it probably does not need to be said, uncompleted, i.e. busy or unanswered calls do not count.

As the world moves toward international long distance resale, this antiquated settlement procedure will probably be replaced by a cost oriented settlement procedure. Most PTTs steadfastly refuse to even discuss changing the current settlement process, because doing so would acknowledge the inevitability of competition. Callback, itself, could have an impact on changing the settlement process and could accelerate the advent of global competition.

Chapter 26
IP Specific Issues

VoIP introduces some real complications into the otherwise well-structured telecom world. In many ways, it contains the best of all worlds — low cost, high scalability, and enough confusion as to legal issues so that it allows practitioners a lot of "wiggle room." The enthusiasm over the Internet and its unseen possible positive benefits causes a *laisez faire* attitude to be taken by many. This hands-off approach arises out of a generally held belief that the Internet is a rebirth of the Industrial Revolution, and thus it must be preserved and fostered at any cost.

Consequently, many governing bodies adopted an official policy of non-interference. Most notable was the European Union, who officially decided to not even consider any significant Internet regulation for several years. This liberal stance towards the Internet, coupled with its attractive bandwidth, led many would-be competitors to focus their attentions on the Internet.

Callback, and other gray market competitive initiatives, had progressed about as far as they could go when the Internet began its meteoric rise. During this time, some callback providers had grown to the point that they could see the need for, and could finance the building of, private network facilities to carry their increasing volume of traffic.

Yet, at the same time, these companies were in a price war; fighting a vicious battle where price was the most important single factor. They had gotten themselves into this unenviable position through unsophisticated sales efforts, often using untrained and unpaid agents to sell their products. This accelerated the march towards commodity competition, squeezing profits, but building volume.

There was a simple technological solution to this dilemma. Since callback was using tariffed telecom services, and playing an arbitrage game, they were still locked into paying high per minute carrier rates. The cost of dedicated facilities was much less, and easier to control and manage.

The only problem was that there were very few places you could put in dedicated lines to provide such service. Satellite communications had a certain appeal, but two factors made it a short term strategy. First was the delay. Satellites in geosynchronous orbits have a noticeable delay that makes them less attractive. If the other service is satellite (which is diminishing in use for voice communications), it can be an alternative, but if terrestrial circuits, like fiber optics are available, the satellite circuits have a competitive disadvantage.

The second problem is that satellite facilities involve a sizable antenna structure, and are, therefore, very noticeable. If the service is legally questionable, it is difficult to hide. Complicating this even more is that you have to connect to the PSTN somewhere, and ordering a number of connections to a building with an 18 foot satellite dish on top can draw unwanted attention from the very source of the PSTN lines.

Against this backdrop, VoIP entered the scene. It had the technical capability of carrying the traffic, and since it involved the Internet, it was often undetected, or ignored. If the partner in the foreign location had access to considerable Internet bandwidth, and could get access to PSTN connections, the technological solution became obvious.

It did not take telecom entrepreneurs long to figure out that the perfect business partner was embodied in the growing body of Internet Service Providers springing up around the world. They had everything that was needed to assure a quick start — compatible technical facilities and personnel, Internet bandwidth and the ability to order PSTN facilities without raising too many eyebrows.

Whether they sought out existing ISPs, or created their own operation was immaterial, because a regulatory solution to a business problem had been found.

The race was on.

In a few short years, the market for VoIP equipment skyrocketed, and companies like Nortel (an AT&T Lucent spinoff) and Cisco were profiting from the sale of products designed to carry voice traffic over the Internet. This is not to say that the regulatory pressures went away in a heartbeat, in fact they still exist today, but the attitude toward rigid regulation has changed as the realization set in that trying to suppress growth is probably a losing battle.

Just as callback served as a catalyst in the opening of telecom markets around the world, VoIP (in many cases) swung the door wide open. Prices for calls on certain routes fell 95% in some instances. China, once one of the most expensive destinations, has now joined the ranks of the sub-5 cent destinations. This is in light of the fact that China has among the lowest penetration of telephones in the world and has joined other countries in making VoIP illegal. But unlike others, China is actually attempting to enforce the regulation.

The attempts at regulating or stopping VoIP are vaguely reminiscent of similar efforts by AT&T to stop competition in the 1960s. Callback and VoIP both use perfectly legal services, for the most part. The complaint that the local regulatory authorities have is in the *use* of the perfectly legal services, just as with the Carterphone decision. This complicates enforcement considerably in most free countries, although a lack of enforcement effort or success does not make the use legal, *per se*.

This leads to the problem that is characteristic of VoIP, as well as with many other alternative services that are not fully recognized and approved. That is, they are subject to interruption without notice — there are ways that PTTs can detect when VoIP is being used, and so a PTT may take action from time to time. If an ISP, who offers dial-up services, begins to *originate* a great deal of traffic, as they would if they started terminating traffic into their PSTN, the PTT could see this and act, causing the circuit to go high and dry.

This "here today — gone tomorrow" aspect to VoIP has some significant impacts, besides the obvious. For one, it forces anyone who is buying VoIP services, whether it is for their own use or resale, to make sure that they have a "hot" alternative route that can be brought up on short, i.e. immediate, notice. Of course, this type of diversification isn't a bad idea, but in the past the carrier took care of this for its users. Today, the users will have to think out and implement their own strategies, and not rely on the carrier to complete each and every connection. This is quite a departure from the past when a user simply sent a call to the telephone company, and could rely on the fact that the call would go through.

The ultimate price of free market competition for the consumer is the "right" to make wrong choices and to suffer accordingly. In the grand scheme of things in the new telecom world order, the regulatory regime will have a great influence on how and when these changes occur.

For VoIP specifically, watch for the developments as they unfold, it will undoubtedly be an interesting battle!

Chapter 27
The ITU, and Other Modern Dinosaurs

Ah, the perennial "they". The irrefutable source of knowledge. The blame for all things that go wrong. The nameless, faceless perpetrator of all evil.

In this case, though, "they" have a name — the International Telecommunications Union. They have a face — the permanent United Nations agency given the broad and rather abstract task of developing worldwide telecommunications according to their charter. "They" are the good ol' boy network of incumbent, mostly government owned, usually monopolistic, national telecommunications carriers. One hundred eighty-seven, more or less, full members who have a vested interest in maintaining the status quo in telecommunications, regardless of the cost in inflated service charges, lost opportunity, or technological innovation.

They have one agenda, to promulgate the antiquated system of bilateral agreements that set the baseline for telecommunications charges, regardless of the actual cost involved. So they can hold onto the turf they have ruled over, and profited from, over the decades. Until recently the ITU was silent, only occasionally commenting on the most esoteric of technical issues, modem signaling standards, telecommunications network interface standards, etc. But behind the scenes, the ITU worked very hard to frustrate the efforts of anyone brash enough to challenge the status quo, or to suggest that telecommunications services should be competitively offered or to let free market economics enter into the equation for setting the cost of telecommunications services.

It is remarkable, and probably due more to a lapse in attention to detail than anything else, that the ITU allowed Recommendation H.323 to come out at all. And to compound the error, they permitted Versions 2, 3 and 4 to build on H.323 and enhance VoIP capabilities to describe a functional interface to the PSTN. Maybe they didn't know what they were unleashing, or maybe they didn't believe that VoIP would grow the way it has. But, regardless of the reasons, the ITU made a material contribution to the world of emerging carriers by its introduction of H.323.

The significance of this contribution cannot be underestimated. Without some acknowledgment to a *de facto* standard by some influential standards body, the manufacturers of VoIP and emerging telecom equipment might have battled for market dominance for years, stifling the new technologies in the process.

It is an odd phenomenon that in the early days of any new technology, the participants will fight so hard for dominance, while refusing to concede anything. So much so that they may end up killing the exact thing that they are trying to promote. Worse case, someone else may act like Solomon and split the baby for them. In this case, the ITU acted with rare leadership. It put the interests of the industry above the selfish interests of the chose few and made a decisive move.

It was totally out of character, especially when based on the past stances taken on competition in general, and callback in particular. This is the aspect that is so intriguing, and why we appear to be obsessing on it. If, and that is a big IF, this represents a change in the attitude of the ITU, it could signal an even more significant change in the minds of its constituents.

A truly earth-moving change; but not completely without logic.

Many of the world's biggest and most entrenched telecom companies, the former PTTs and the monopolists, are now seeking opportunities outside their own borders. One notable example of this is Singapore Telecom. Singapore is a city state in central Asia, with a population of around four million people. It is home to an enormously successful business community and its racially and culturally diverse populace enjoys one of the highest standards of living in the world. This city state also has one of the most modern telecommunications systems in the world. One that's well financed, well engineered, and well run.

Singapore Telecom is an example of what could have been a dinosaur telecom, but instead it decided to move forward. Although it sports a monopolistic PTT history, its history also indicates Singapore Telecom has always been a farsighted entrepreneurial enterprise. This has allowed it to not just prosper, but to grow and expand. True, the company has fought, and fought hard, to sustain its monopoly at home. But, at exactly the same moment that the PTT was fighting to curtail activities like callback (*The International CallBack Book* was once banned in Singapore!) it was investing, and investing heavily, in competitive activities outside its own borders. In fact, in the mid-90s, Singapore Telecom bought a piece of a US-based callback company, and participated in the provisioning of callback services! Even more remarkable was that the telco was actively involved in providing and reselling callback services in Singapore.

This pattern is likely to continue throughout the world. We have seen it in Europe. There, once the telecom environment was liberalized and opened to more free market competition, carriers began competing with each other — France Telecom and Duetsche Telekom had a very good thing going, working together in all sorts of global collaborations. That is until DT one day decided to step on some ground that FT believed belonged to them. Now they are rivals once again.

🔖 Chapter 28

Multilateralism, the WTO and Other Pipe Dreams

The traditional structure of the telecom industry was a series of bilateral agreements between parties. If there are 238 tele-jurisdictions, then logically, there could be over 56,000 of these agreements. In fact, that is exactly what did occur in the past under the age old CCITT (which is now the ITU) structure. Once alternate carriers began to emerge, first in the US, then around the world, the number of possibilities grew geometrically.

Some countries gave their PTT the sole responsibility for negotiating and maintaining these structures. The US set up a regulatory system so that no US carrier could get preferential rates over another US carrier. This, however, typically only applied to the largest, Tier 1 carriers, if at all. Certainly, no one at the FCC was charged with actively seeking violations of this rule. Had they done so in the 1990s, they would have been a very busy indeed.

Although there had been talk for years about a multilateral structure for tariffs and trade negotiations, the world was not ready for what was beginning to happen in telecommunications. Yet, bilateral agreements were an issue in all manners of international trade. On the heels of World War II, a movement began to generalize trade agreements and contracts among nations and, thus, spur trade. This movement was formalized to GATT (General Agreement on Tariffs and Trade), which started with the Allied nations, but spread over the years to include other countries around the world.

In 1986, the Uruguay round of GATT promised to open world markets, end protectionism and thus improve the global economy. This was an ambitious idea, and it took eight years before the World Trade Organization (WTO) was created in 1994 to oversee such an effort. The concept behind the WTO was to allow participating nations to create a formal agreement pool, where one offer, called a "concession," could be entered by a nation. The concession would be available to participating nations that could meet the terms without the need for a separate bilateral agreement.

In 1996, the WTO concluded a landmark Fourth Protocol, under which member nations would open their telecom markets using the WTO GATT principles. This was accepted by the vast majority of the nations representing the world's largest populations, but most notably sans the world's largest country, China. This agreement went into place, after a short delay, in 1998, a few months after the EU mandated opening of the European telecom market.

While these developments are encouraging, and represent formal market openings, in reality, instead of creating a telecom democracy, they create an aristocracy, opening markets to only the anointed players. The minimum criteria for formal entry into most markets are onerous. For instance in Germany, a license to operate usually costs more than US$5 million and requires an initial investment of switching and network facilities that only the most affluent and well financed competitors can afford.

Telecom startups are undaunted. Even if the reality is still a ways off, they are encouraged by the prospect of competition. This formal recognition of *some* competition also pique the interest of users, big and small, who see this development as condoning alternate carriers and legitimizing the entire shadow industry.

VoIP operators used this opportunity to step in and offer highly sophisticated services, even if they had not achieved full legal operating authority.

Chapter 29
The End of
Global Regulation

All of these forces, formal and "unauthorized", have added a lot of competition, all of which put great pressure on prices. In the year 2000, we saw a big shakeout among the telcos, large and small. Not only did the biggest carriers suffer, including AT&T, MCI/Worldcom, British Telecom, but many of the high flying startups — Star Telecom, World Access, Viatel, Ursus, RSL Com and other well-financed, well managed enterprises were on the brink of extinction. These companies were joined by a host of equipment manufacturers — the mighty Lucent was teetering on the verge of becoming a penny stock, and was an oft-mentioned takeover target. Even the wireless companies, considered by most to be invulnerable, were at risk.

All of the old rules went out the window.

Survival, once thought virtually assured, became the main topic of discussion in boardrooms around the world. One could imagine the Chairman of the Board of a global, multinational Top 1000 telecommunications conglomerate, turning the lights in the boardroom to ruby red, donning a battle helmet, sucking on a cheap cigar, and, with a glazed look in his eyes, saying, "It's do or die!"

Investors, big and small, were thinking the unthinkable. It was not only possible to lose significant sums of money on stocks that had once been an iron clad guarantee, but they had already had the experience. Lucent, the former equipment arm of AT&T, which had been over $60 a share in the Fall of 2000, was close to $5 in the Spring of 2001. This meant that it had lost over 90% of its value in a period of less than 6 months. Before Lucent was spun off, AT&T had been the most widely held stock in the world, thus Lucent was extensively held.

It wasn't just the Internet companies, either. Everyone expected that there would be an eventual shakeout in the Internet community that was spawned in the late 90s. These companies were seen as risky, so the possibility of a loss was not unexpected. But was it possible for telecommunications stocks, many of which were owned or partly owned by the government itself? Unthinkable! Besides, wasn't telecommunications a *basic* industry? Like food, fuel and housing?

Well, times have changed. What has not changed is the long term outlook for telecommunications. Nobody is questioning whether the industry will survive. It will. What is being questioned is which companies will be among the survivors. Some of the old favorites are no longer. For investors, the message is obvious, hedge your bet, and don't put too much into any one company, because some will fail. No company is immune from the risk. Lucent is as likely to fail, as is a startup dotcom company.

This new paradigm has regulatory impact as well. The protectionism of the past was a safe bet. Whatever the makeup of the regulatory authority, it was clear the market leaders would stay in that position, and the role of the regulator was to control the rollout of competition. It was not possible to make a mistake with any consequences. If you allowed too little competition, things did not change fast, and the incumbents would maintain their positions of leadership. If you allowed too much competition, the competitors would capture a larger percentage of market share and grow faster.

Because of the enormous growth rate in the telecommunications industry, in many cases the new competitors were only capturing a portion of the *growth* of telecommunications. Even in the "bat out of hell" growth in telecommunications competition in the US during its market opening, AT&T 's annual revenues NEVER went down, not once. They lost 40% of the market, from 100% in 1968, to under 60% in the 90s, without ever losing revenue. In fact, the revenue growth was strong throughout its losing run.

This is even more remarkable in light of the fact that unit prices for telecommunications services were dropping rapidly during the same period of time, due to the competitive pressures that were brought to bear. The average long distance minute in the US went from 35 cents in the early 80s to less than 10 cents in less than 15 years, a drop of over 60%.

How was it possible to lose 60% of the market, drop prices 60%, and still maintain a strong revenue growth, at impressive margins? (Until the late 90s, AT&T had NEVER missed a dividend to its stockholders.) Simple, the growth of demand for telecommunications products and services made up for the change in competitive structure and pricing pressures. A regulator, therefore, could never make a wrong decision; there was simply no downside to any decision — no unforeseen potholes in the road. No way to make a serious mistake. If you were too liberal, competition would simply get a higher percentage of the growth rate.

But things have changed. The unthinkable is now thinkable, even probable. A major primary telecommunications supplier could fail. If a new competitor failed, well, that could be tolerated. But, a primary telecommunications company, AT&T, BT, or any of the over 150 PTTs, not to mention the hoards of LECs, to actually FAIL? Impossible!

You better believe it can happen — and, sadly, it probably will.

Now, faced with the prospect that they COULD make a serious mistake, regulators are faced with a dilemma, how do you assure the survival of the incumbent, and still hedge your bet enough to allow for a few strategic competitors to make the final cut? This assumes, of course, that the regulators have sufficient power to make these decisions. We have mentioned before that much of the new competitive telecommunications business is not subject to rigorous regulatory review and oversight, so the influence of the regulator is almost a moot point. Unless, of course, the regulator also intends to be an enforcer of the policy it sets.

In most countries, regulatory enforcement is separate from policy setting, and often only performed in reaction to formal complaints or actions initiated elsewhere. Carrying unauthorized VoIP traffic over the Internet into a country is usually not treated with the same gravity as drug smuggling. The regulation is usually not proactive, and only occurs when there is sufficient public interest involved. There is also the issue of resources. No matter how motivated a regulator may be to enforce a particular policy, it may not have the resources and assets at hand to perform the enforcement.

There are no "phone cops" at the FCC. I have never heard of an FCC inspector marching into a collocation hotel, shouting "book 'em, Dano!" If this were the case, 60 Hudson Street in New York, 1 Wiltshire in Los Angeles, 1101 Biscayne Boulevard in Miami and dozens of other US Points of Presence, there would be battalions of enforcers in these locations, and dozens of other equally "hot" spots around the country. But, this is not the case, and it is not planned. To the best of my knowledge, there has never even been any serious discussion of any such enforcement. Even if there were, there are not enough lawyers and other administrative personnel to fully prosecute such an action.

Would they find blatantly illegal connections and service arrangements? What do you think? There are enough illegal telecommunications services being provisioned to fill several careers with legal actions.

Why aren't these laws being enforced? The fact is that assets are so diversified that busting one or even a few dozen of the gray market providers would barely make a scratch in the surface. So, an enforcement process is unlikely to produce any concrete results.

The other consideration is that the regulator faces a couple of new realities today. Beyond the enforcement reality, is the fact that a significant portion of the low cost, gray market VoIP minutes end up in the hands of primary carriers, who launder them, and integrate them into their networks. They are able to cream skim, and purchase only high quality, high availability minutes, so network quality overall does not suffer. Also, because of the way these minutes move between carriers, providers and clearinghouses, it is impossible for anyone to ascertain where the underlying minutes were created. Even AT&T, in 1999, began a concerted effort to buy such minutes, working through brokers and consultants. A special division was established to facilitate buying gray market minutes and a serious effort is ongoing today. The contractual arrangements with AT&T contain enough confidentiality so that AT&T can maintain a reasonable level of plausible deniability.

If a PTT, or a foreign correspondent of AT&T could ever prove that AT&T was using gray market minutes, it could take action. So AT&T makes sure that it is sufficiently insulated to be able to keep its plausible deniability intact. AT&T is not the first carrier to make such arrangements. It's rumored that other Tier 1 carriers have been involved in the practice for a long time. Smaller carriers, like Primus, have always been involved and, in fact, built a substantial portion of its business on providing low cost, high quality international connections.

Aggressive enforcement of gray market minutes could significantly dampen the availability of minutes — which may apply in any country anywhere! Best case, since many primary, as well as alternative carriers, in every nearly country are using these minutes, they would not be likely to support such a move. Worst case, they may actually be supplying some of these minutes, and would not appreciate seeing their markets dry up.

It would appear that the Genie has been let out of the bottle, and that we have just taken another step in the transformation from a tightly regulated telecommunications environment, to a more free market, competitive industry. Perhaps visionaries at the FCC and the EU realized what was about to happen, and this is why they delayed taking any action on VoIP. It is unusual for government to restrain taking action for altruistic reasons, so maybe they realized that, if they delayed taking a controversial action on an inevitable development, they could avoid an embarrassing defeat.

If this does represent a new direction in telecommunications, it has vast implications for regulators and regulated alike. It means that less regulation will be the wave of the future, and those jurisdictions that attempt to buck the trend may be overwhelmed by it. Remember the actions of the farsighted management at Singapore Telecom. Maybe their duplicitous effort was not as schizophrenic as it appeared. Maybe they saw what was coming, and didn't want to alert the world to it, so they gave the appearance of maintaining the status quo, but intended to be a leader in the new telecom world all along.

Maybe the future of telecommunications regulation will be distinguished by those of the old guard who successfully transition to the new realities and prosper; and those among the ancient regime that don't or can't might disappear into the abyss of the past.

Section V

Business Strategy

Chapter 30
Running a
Telecom Business

In many ways, the VoIP entrepreneur or new generation telecommunications business is very similar to running any other telecommunications business. The key differences are the international aspect, the technological sophistication of the equipment and systems involved and the sheer volumes of money and resources that are required. Just like any business, the VoIP entrepreneur needs to find and recruit the proper personnel, arrange financing for operations and equipment, collect bills and money owed, and plan and execute a well thought out term strategy for the business.

Sales

The most crucial element for success in a telecommunications enterprise is the ability to sell and distribute product. In the case of VoIP, one of the most challenging is to arrive at a cost effective sales and distribution strategy. The best advice is that there are three keys to be successful — sales, sales, and sales. If all else fails, if you have sales, you can work through the others. Conversely, if you do not have sales, the rest can be flawless, and you will fail anyway.

An effective sales strategy alone does not guarantee success, but it does facilitate the way to success. It is an essential ingredient, in that without it, you cannot ever hope to be successful, but it doesn't guarantee success by itself.

Sales Channels

There are many theories on how to establish an effective sales organization. Much practical experience can be gleaned from other people who have started, not only a VoIP business, but have experience with other businesses as well. Selling VoIP services is a bit more complicated, especially if the marketing of advanced services is involved. Add to this the increasingly international character of telecommunications, and you have a picture that differs quite a lot from the typical utility service.

Telecommunications has traditionally been mass marketed on the low end, with extensive use of advertising in mass media. At the high end of the market, direct sales efforts have been the norm. The very lucrative middle ground has a history of bipolar efforts, an eclectic mixture of over-attention and total ignorance. The biggest struggle has been between direct sales channels and indirect sales efforts. In fact, a case can be made that competition in the US was kick started by an indirect sales strategy created by AT&T, affectionately know as Tariff 12.

This reference to the FCC filed tariff number also testifies to the struggle going on in the US between regulation and competition. In the 80s, AT&T felt hampered by continuing FCC regulation of its operations, and the relatively free hand that most of its competitors had. So, an innovative plan was devised that would allow it to direct mid and lower market sales through a network of sales agents, called "aggregators", who would sell AT&T services, but without the cumbersome regulation that AT&T faced.

We are probably being too generous to AT&T in saying that it devised this plan. It actually happened, over AT&T's objection, in great measure. Earlier, it had created a tariff that allowed its larger, multi-location customers to aggregate their services to maximize discounts. These discounts were very important to AT&T in fighting off the competitive advances of MCI, Sprint and others, who were attacking AT&T's key customers.

A loophole in the tariff allowed very loosely associated companies to be aggregated, so some creative entrepreneurs began signing Tariff 12s and providing premium AT&T services to customers who were really not allied in any meaningful way. AT&T made a perfunctory gesture toward refusing these "unwanted" interveners, although the effort was destined to fail because of the deliberately vague requirements it set for determining if a company was really aligned with the Tariff 12 holder. So, when it became obvious that the

FCC was not going to bail out AT&T, it decided to save face, and claim that Tariff 12 was instead a clever strategy to stop competitive erosion of middle and low markets.

Although the transparent public relations effort at spinning this regulatory defeat failed, Tariff 12 prospered. It seemed to be a win-win-win situation, since AT&T thought it got a powerful competitive tool, the "aggregators" made a decent profit with little or no investment, and the customers got some relief from their long distance bills.

The problem was that aggregators found it much easier to sell a slight reduction in AT&T rates to existing AT&T customers who would probably not have switched anyway, than to those who had already accepted much deeper discounts from AT&T's competitors. The way AT&T viewed it was that it added nothing to its competitive counter-efforts, and only took away revenue it would have gotten anyway. This "cream skimming", as it was called, became a serious bone of contention, and caused the already tenuous relationships with the growing troops of aggregators to degenerate into outright war. AT&T's former "network of marketing partners", the aggre**ga**tors, were now called "aggre**va**tors" and were dreaded, hated and fought, tooth and nail, with every dirty trick in the book. This war resulted in years of litigation, making many lawyers, and a few aggregators, very rich.

This dilemma in alternate sales channels is not unique to telecommunications, but it has been especially acute in the rollout of competitive telecommunications services around the world. The trade offs are extremes. Using an indirect sales channel to roll out sales efforts gives the company a fast start, obviates the need for sales related expenses and can create an instant presence. On the other hand, the commissions will usually be higher than what would be paid to employees, and more importantly, the customers would be owned by the agent, written agreements notwithstanding.

Let's examine the two basic types of sales distribution — direct channels and indirect channels. We will also look at some of the considerations to use in determining which one is the best for your business. One of the most exciting things about this industry is the fact that it is so very new and many creative ideas have yet to be tried, so don't restrict your thinking, especially in marketing.

Indirect Channels
Probably the predominant method of marketing VoIP and enhanced telecommunications services today is through indirect channels. An indirect channel,

by definition, is one in which the sales resources do not report directly to the service provider. In other words, there is a business-to-business or indirect relationship between the service provider and the agent.

This has several implications.

One is that the agent or sales channel is independent of the service provider and can carry other products, sometimes even represent other competitive companies. These agents often move from one company to another within the same industry. It also means that the sale representative is not a direct employee of the service provider; but is either an employee of a company with an indirect relationship or an independent agent.

In a startup callback company, there are some notable advantages to this kind of a scheme. The most obvious one, of course, is that you do not have to recruit, hire or pay sales agents. This saves a lot of time and money, especially if you can find sales agents or distribution channels that already exist with a customer base.

On an ongoing basis, one of the big advantages is that these arrangements are usually set up on a commission scale. The amount of commissions paid and, therefore, the sales expense is directly proportionate to the level of sales. So in the beginning, when the level of sales is low, the direct sales expense is

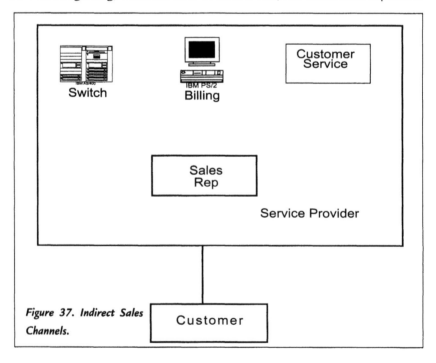

Figure 37. Indirect Sales Channels.

also proportionately low. As the level of sales increases, the amount of money paid for direct sales expense increases, but in a very predictable manner. Other than the usual startup expenses for producing sales materials, brochures, etc., there is not a big initial investment in obtaining an indirect sales force.

Some of the disadvantages to this scheme are that since it is an indirect relationship and the agent could go to other products or even other companies, it provides very little protection to you as a service provider in maintaining these resources. Another problem is that generally an indirect sales force is less expensive for a high level of sales than a direct sales force. Direct sales forces are usually paid on the basis of a fixed salary plus a commission. Therefore, in the beginning, the percent of sales costs will be somewhat high because for the first dollar of revenue, you will have already put out the sales expense.

It is also a reality that whatever sales resource you put into place (for instance, a new employee), it will take some time before that new resource actually starts generating revenue. This is especially true if the sales resource is going to be cold calling or does not have a list of prospects from which to begin work. At any rate, there will be some time lag from the time you start paying until the resource actually starts to produce revenue. In an indirect relationship, since it is based on a commission, the only investment you will have is the expense of locating the agent or company you want. Of course, this, in itself, can take some time, but it generally will produce bigger results because the time spent in recruiting a company with 30 sales agents, who will be selling your product, pays off a lot better than chasing one customer.

A few of the disadvantages of an indirect sales force have to do with customer service. If you are not careful in structuring the arrangement, the sales agents will tend to want to make the sale and walk away because they are not paid a commission for customer service. You can avoid this by tying future commissions to customer satisfaction. In other words, the sales people do not make all of their commissions up front. It is paid monthly and a portion of it is held back for a period of time to assure that the customer is ultimately satisfied. If you want to build a long term relationship, of course, you will continue to pay the agents a residual, albeit a lot smaller than the initial amount, for maintaining the customer, and therefore, the revenue stream, over an extended period of time.

Many companies pay their agents based on the level of collections in a given month, instead of sales. Paying on collection has several advantages. First, it keeps the sales agent on the hook for maintaining customer satisfaction, at least until the bill is paid. It also makes sure that all the business claimed by the agent is good business. Last, but not least, it helps to control your expenses because until the bill is collected, you do not pay a commission to the agent.

Another disadvantage with dealing through agents is that, because you are typically dealing across international boundaries, there are logistic and legal problems. A logistic problem, of course, is maintaining a high level of contact with the employees. It is not easy to have a face-to-face meeting every Friday morning if you are in one country and the agents are in the other country. What you have to do is substitute quality for quantity. Make sure that if you have agents in other countries that they give you periodic reports on what they are doing and that you have regularly scheduled telephone conferences with them to go over their sales projections and their actual results.

There are many places and many ways for finding successful sales agents. Many people go out and try to recruit people who have sold in other areas and are looking for something to do, sort of an entrepreneurial type, to sell their callback service. Although there are some advantages to this, some disadvantages are that you don't necessarily find sales agents that have existing customer relationships that will result in a quick turn of revenue. For VoIP, this makes the ISP relationship even more valuable, since the ISP can lead the sales effort in addition to providing the technical services.

The other category of sale agents you can look for are companies or people who are currently selling to an existing base of customers in a related field. For example, someone who sells PBX or telephone equipment is a natural adjunct to selling telecommunications services. In fact, this can be a very effective way to market telecommunications because providing long distance service is a natural extension to providing the telephone equipment. Sometimes it can be a very minor point for a customer to sign up for a service at the same time that they buy new telephone equipment.

Direct Sales Channels

In the other category of sales channel, the direct sales channel, you develop your own, paid and commissioned sales force. This is probably the more attractive proposition for most companies because of the level of control over the resource that it provides. In an indirect relationship where the agents may also sell or provide other products or services, you have to compete for the attention of that sales force to sell your product. This means you may have to provide incentives, spiffs, or bonuses to them to entice them to sell your product over the other products that they are carrying.

With a direct sales force, since they are basically employees of your company, you have complete control to push the products you want them to push. The disadvantage, of course, of a direct sales force is that you must make a significant investment long before the first dollar of revenue comes in. An offsetting advantage to this is that generally, as time goes on, you can raise, control and restrain the sales commission percentage that you are paying. Savvy sales representatives understand that when they get into a new field or area, the commission rate will be very high, but as the field or area matures and becomes more accepted, those commissions will not be as lucrative.

The disadvantages of a direct sales channel can be onerous. There is, of course, the issue of a corporation hiring employees. This means that you must collect taxes, and other fees associated with employees. You also face transportation and travel expenses. There are also regulations and restrictions on hiring, firing and managing these employees. And you find that vacations, holidays and sick days, are, indeed, a COST to your business.

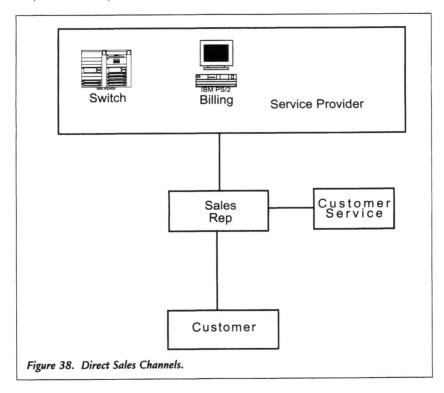

Figure 38. Direct Sales Channels.

This is going to increase your administrative costs considerably because you need to make sure that you not only meet all the US requirements for a US employer, but you also have to meet the requirements for employers in other countries as well. Many countries have very stringent laws on how foreign corporations can deal with its citizens. For example, there are countries where it is illegal to fire an employee without government authorization and notification. The complication of doing business in those countries with direct employees is obvious.

In reality, you should probably use a combination of direct and indirect channels. This can serve to blend the best of both worlds, and make sure that you have good market coverage. And, don't be afraid of a little conflict between your sales channels, it can be very healthy, and can also assure that you are not leaving any stone unturned. Trying too hard to avoid overlap almost always assures that you leave gaps in your coverage.

Marketing Strategies

There are many entrepreneurs with the resources and desire to establish a telecommunications company. There are several key aspects to doing this, and several important disciplines are involved. The ability to set up and run a company is one of the basic skills necessary in this, or any other, business venture. It is difficult to learn the fundamentals of business management, while at the same time, learning the technical operations management of a complex communications system, as well as getting your feet into the world of marketing. If you lack one or more of the critical skills you should consider finding partners or managers who are proven to be skillful in the areas where you are weak. This is not an industry where amateurs can easily succeed!

The first step in setting up your company, before you make any critical decisions, is to decide on the who, what, why and how of your business plan. Many telecommunications services tend to try to be all things to all people, and don't target their service to any particular market segment. Large, well-established companies may be able to get away with such a "shotgun" approach to sales, but a small, new company courts disaster if it doesn't have a clear focus. You can waste a lot of scarce resources by trying to grab at anything — sometimes referred to as the "ready, fire, aim" approach.

When you have a marketing plan that is focused on a group of consumers, and have determined the viability of the marketing plan, the rest of the decisions, while still not easy, at least have a point of reference by which the other decisions can be made.

Promoting the Service

Once you have commenced operations, and have a working distribution channel, you are ready to begin promoting and selling the service. What is the best approach? You could put advertisements in the newspapers where you are intending to sell the service, or you could launch a cold calling campaign, or you might begin forming strategic relationships with business partners at home and overseas. Then, of course, there is always the Internet.

All of these have their application, but first you must go back to the basics. If you have written and followed a business plan, you will have a very clear concept of what you are selling and to whom. Now you must make sure that you have in place the "stuff" that your sales agents and partners will need to begin putting paying customers on the service. Without them, the rest of what you have done is completely useless.

There are some basics that you will find necessary, no matter who your target audience might be. Our purpose here is to help identify what you may need and what it should contain, but the details will be up to you, your business plan, and what your selling resources are telling you that they need.

It is clear that you will need some material for your pre-sales effort, and others for the post sale customer servicing.

Pre-Sales Materials

The physical format of the pre-sales materials will be highly dependent upon the distribution channels you intend to utilize. If the majority of your sales will be face to face, you will probably need four-color brochures, some form of stand up presentation and possibly a written proposal. As part of this, you will also need many different types of prospecting and follow up letters and other sales correspondence. Your sales effort will also necessitate some credit applications, agreements, implementation plans and customer instruction information.

If your sales effort is more oriented to telephone and/or long distance contacts, you need sample scripts and fax oriented materials as well. The importance of a clear business plan cannot be overemphasized. We have witnessed employee training sessions that completely fell apart because the person doing the training did not understand the details of the transaction itself.

First, define the steps in the sales process; the requirements for each step will become fairly obvious. Step through the process with a business associate and see what it looks and feels like. You will be able to make a fairly educated estimation as to what you need if you do this. The person playing the customer in this dry run should also ask the questions they feel they need answered.

Also investigate the types of media you will need. Much international business is done via telephone and fax, so you need to make sure that you have the materials in the proper format for your customers. If you are going to be using agents or other face to face selling methods, you might want to have some of the materials printed professionally for their use.

Here is a checklist of the sales materials you should have on hand.

✔ A descriptive brochure outlining the service or services you offer. Be sure to highlight any services or features that you have that set you apart from your competition. Don't leave anything out, make sure that you specify what you do, even if it seems obvious to you, or is something that you are sure your competition has. You still need to tell prospects that you have it too.

✔ A pricing information sheet, whether it is exemplary pricing or examples of common types of calls, the user will need some indication of how much they will pay for calls. Give this document a lot of thought — you want to present a clear, concise view of your rate structure since many potential customers' focus will be on cost.

✔ A sheet with complete descriptions of the services you offer. Thoroughly describe the services offered. Saying "complete voice mail' is not enough information for the user to determine if the voice mail meets his or her needs. The user may not have any experience with voice mail, so they will not understand what "complete" means. This is probably the most neglected area of any service offering. Those selling technology and not applications will pay for their sins heavily when they do not describe the service in agonizing detail to prospective customers. This is one time that it is all right to treat your customers like complete idiots. They will love you for it.

✔ A thorough explanation of your credit requirements is important. Credit requirements can be one of the touchiest issues in the relationship, so it is best to spell out what you expect. Be sure to mention all the options and details. You may find that the customer has already selected the way they want to secure the account before they call.

✔ A customer application, ready to fill out and mail or fax is a necessity. Make this form very simple, and avoid any extraneous or marketing research type

questions. The easier it is for the customer to complete and fax the application, the more likely they are to sign up for the service. If you need credit information, or intend to use a credit card merchant account, be sure to
get a signature, and use the prescribed credit card statement.

✔ Provide background information on alternate telecommunications service
providers. Many people do not understand what this is all about, and will
naturally feel a bit uneasy about changing. Tell them what you have to
offer, how you are able to save money for the user, and what they can
expect from your service. You should be very honest about the disadvantages as well, but do not dwell on them.

✔ Provide background information on your company. The prospective customer will want to have some idea of who you are, how long you have
been in business, and some details about your company. You are asking
them for some details about them or their company, especially if you are
going to ask for some form of security, so tell them a bit about yourself.
They must have a level of trust in you to do business. This is your opportunity to give them the information necessary for such trust.

Pricing Strategies

Prices for competitive telecom services have been in a free fall for over 15 years,
no matter which service you look at — long distance, wireless, Internet access
— anything that has been offered competitively. Notably absent from the
downward pressure of competition are local services. There are many reasons
for the lack of any significant level of local competition, but the result is the
same — HIGHER local prices. If you don't believe it, add up your telephone bill
for local services today, and compare it to what it was 10 or 15 years ago.
Again, we are going to use the US as the example, since competition has been
a factor of life here much longer and more intensely than in other places.

Some of the increase is due to taxes — Universal Service Fund, Access
Charges and the never-ending Federal Excise Tax. (The official position is that
USF and Access Charges are not taxes, but take a look at the dictionary definition of "tax" to see if they are or not. Nevertheless, they offer no additional
value to the user, and cannot be avoided.) Some of the increases have come
in the form of new, "necessary" services. Granted, the majority only offer
greater functionality, but their almost universal acceptance makes them nearly a necessity — most notably, call forwarding, call waiting and Caller ID.

Local telephone companies can say all they want about how the cost of local service has gone down relative to the cost of a loaf of bread, but the fact remains that monopoly local service has held its price, while competitive services have lowered prices. Now do you see why the Local Exchange Carriers fight so hard to protect this turf?

So, telecom companies, other than those fortunate enough to offer non-competitive services, will find that pricing products and services is one of the most important things that they do. It does more than just make a billable event a reality, it establishes the position in the marketplace and defines the group of customers that the company will do business with. It also determines how many customers will be enticed to do business. Simplistically speaking, if the price is set too high, not enough customers will be attracted to the service. If set too low, profit opportunities will be missed.

There is a great temptation to go too far one way or the other. A very low price may look like a "silver bullet" against the competition, where a very high price seems to make more margin on lower volume. The correct price will depend on the type of service that is offered. A high quality service, giving the user very low risk, can command a premium price. A lower price will expedite sales, and make revenue flow faster.

On the other end of the rate spectrum are those who offer a basic service and whose main attraction is their very low rates. Of course, there is always a trade-off. The lower the rate that you offer, the less confidence the consumer will have in the product or service. The consumer may perceive your offerings to be less reliable because of the price difference, and in fact, a lack of cash flow may not allow capital to be used for the repair and construction of existing facilities — in telecommunications, perception sometimes IS reality.

Price is also dependent on several other factors, including the risk the company is willing to take in the area of billing and collections. There are several other marketing and operational considerations that affect pricing. These factors need to be considered before intelligent decisions can be made on pricing, whether you are a service provider setting prices or a potential user evaluating them.

Managing Traffic & Carriers

The VoIP business began as an arbitrage business, and despite many efforts to change that, it still remains predominantly an arbitrage business. The great promise of VoIP is to deliver applications, which have great value to the end user, and thus provide enhanced revenues to the service providers. But, the allure of quick revenues via traffic swapping is hard to resist.

Despite some recent progress in the delivery of enhanced applications over VoIP, the actual revenue stream it represents is very small for all but a few service providers. So far, VoIP has largely failed to deliver on the promise of new, killer applications. The future looks promising, but it is still the future. In the meantime we will have to live with the fact that VoIP is primarily an arbitrage business. Even the big VoIP companies, like ITXC, have built their business and their revenues primarily on arbitrage.

Why has arbitrage emerged as the primary application for VoIP?

One of the reasons for this phenomenon is the incredible increase in worldwide traffic occurring as a result of lower unit prices. This is a testimonial to the fact that there is great demand for affordable telecom services all around the world. Even though arbitrage may not be the most attractive application for VoIP, it does pay the bills. It also has a financial advantage — a huge cash flow is generated in a very short period of time. This has the downside of taxing the ability of the VoIP company to actually provide an adequate level of service to its customers.

But arbitrage is the opiate of VoIP.

It provides immediate gratification in the form of an enormous growth in revenue. The only problem with this scenario is that as competitive as the telecommunications business has become, VoIP is just another competitor for consumer telecom expenditures. Margins in this business have slimmed nearly to the point where they no longer exist. At one time, a T-1 full of traffic could easily carry over a quarter million dollars worth of high margin traffic monthly, but today, at a prevailing rate of 5 cents per minute in most of the high volume destinations, a typical T-1 now represent less than five thousand dollars a month in low margin traffic.

There are other issues of concern when VoIP is used primarily for arbitrage. First, the arbitrage business is not very attractive from an investment point of view, despite the seemingly large revenue flows. The already meager resources of most VoIP companies tend to be consumed by the necessities essential to keeping the arbitrage business going, preventing the VoIP company from making higher profits in the more attractive enhanced services arena.

Further, while the cost of VoIP equipment is less than legacy, circuit switched equipment, there is still the necessity of a substantial investment for the infrastructure needed to carry traffic, especially high volume traffic. A high volume of traffic is key to getting enough revenue to attract the attention of investors.

Years ago, when callback was king, international long-distance minutes averaged $1 per minute. Today, VoIP minutes are much lower. Most of the attractive international destinations (including the United States) go for well under 10 cents a minute — some as low as 5 cents per minute. Clearly, as just a bit of math will show you, companies that were once billing $1.00 per minute on the average now must run 10 times as many minutes just to maintain the same revenue flow at 10 cents a minute.

Managing Cash Flow

Of all the skills required in running a telecom business, the most critical very well may be the management of the cash flow. This is a high volume, low margin (relatively speaking) game with lots of opportunities for making mistakes. One misstep could be fatal. Because the nature of the telecommunications business is high volume, there is lots of financial leverage, but if you do not plan carefully, the results could be disastrous.

Credit Policies

Probably no single financial decision is as crucial as the setting of credit policies. This decision impacts nearly every facet of the business, and has the potential to negate an otherwise highly effective effort. Mistakes can be equally damaging, whether the policy is too liberal or too conservative. A serious mistake can also be made in not being consistent, which some companies make in an attempt to be flexible. Every situation may be different, but the standards you use to establish policies should be equally applied. For example, if you decide to relax your deposit requirements to accommodate a big deal, you may be taking on too much risk.

Credit policies are like most business decisions, in that it comes down to an analysis of risk versus opportunity. The principle is simple, the higher the risk, the greater the opportunity to earn. Of course, there is also a greater opportunity to lose. This is also the basic principle of gambling, which credit policies should never resemble.

In telecommunications, losses due to bad credit are probably the single greatest direct cause of failure. You can do everything right — the right product, the right price, a decent margin, good quality, technical and customer support, and a robust billing system — and still fail, due to poor credit practices. Lest you think that we are arguing for strict policies to avoid losses or liberal policies that allow aggressive sales — you're mistaken — actually neither is the answer. The solution to the credit conundrum is to analyze the overall impact on the rest of the business, and to balance this against the risk element. Anytime that you extend a dollar of credit to anyone, you are taking some measure of risk.

Due to the high level of failures in telecommunications in the past few years, credit policies have tightened considerably. These days, almost no one extends credit, insisting on payment up front for services. Nearly every service is guaranteed by something. A credit card, money on deposit, or a guarantee against the local telephone service are the most common forms of guarantee for consumers. For resellers or wholesale accounts, cash deposits or a Letter of Credit are almost always required, and for good reason.

This "cash on the barrel head" requirement has fundamentally and dramatically changed the dynamics of the telecommunications industry. It is no longer possible to start up an operation without a significant infusion of money. You can't do it on someone else's money by tapping into their credit. Companies now need measurable resources to get started.

Bill Collections

Most of the early competitive telecommunications services began as debit systems, especially the unauthorized services. In a debit system, the user pays a certain amount of money before the first call is even made, and as calls are made, the cost of each call is deducted from the balance. The initial payment is usually based on the estimated usage; a typical amount being the estimated usage for a month's service, plus a bit to allow for the payment to be received. If the customer approaches the end of the credit balance before the end of the monthly billing cycle, some sort of mechanism alerts the customer that they are running out of credit. The customer can then take some action to avoid discontinuance of the service, such as send another check, give another authorization to do a credit card transfer or do an electronic funds transfer to the service provider's account. But, if all goes well, and the prepayment estimate is accurate, the customer just receives the bill, and pays it, "replenishing" the debit value.

The key feature of a debit system is that at all points the customer assumes all the risk. The telecommunications company is holding a deposit so that when calls are made, the company has enough cash on deposit to cover the calls being made. The telecommunications companies have no risk because they are always holding customers' money against which they can charge calls.

It is essential, therefore, that the switch is able to calculate the cost of each call, and deduct it immediately. This is called "real time call rating", and without it, the telecommunications company opens itself up to fraud or bad debt. This debit system is the basis of prepaid calling cards, also.

The problem with this approach, from a marketing point of view, is that some of the company's competitors may not require advance payments. The company provides the service and allows the customer to make the telephone calls. After the telephone calls are made, the customer is billed, and has a period of time in which to pay the bill. In this case, the competitor takes the risk that the calls will not be paid. This "credit system" is often employed by the incumbent carriers, who have both a payment history for the customer, and an overall sense for the level of bad debt that will occur. They also have the resources to sustain a predictable level of uncollectibles.

If the incumbent is also providing the local service, they have a high degree of leverage in that they are also providing the basic telephone service to the subscriber. They can force payment of the long distance bill by threatening to cut off all telephone service, eliminating not only long distance calls but local service and in-coming calls as well. This enormous advantage may explain why there is such a heated war being waged over the right of Local Exchange Carriers to provide long distance service.

Competitive telecommunications companies, for the most part, do not have this leverage. Therefore, debit systems are used because they reduce the collection risks of the telecommunications company. Even though the alternative telecommunications rates may be lower, the customer has to put some amount of money up before calls can be made. This is an inconvenience and perceived risk that many customers may not be willing to bear, even for the claimed savings involved.

Telecommunications companies who are trying to attract larger businesses may find it necessary to advance credit to these companies in order to win the business. Many, especially the larger companies, are not used to a debit type of business, so approval for this type of expenditure may be difficult or impossible to attain. Therefore, the competitive telecommunications company will find itself faced with a new situation if it begins supplying services to enterprise customers. At this point, the financial game will have to change and there will be some level of uncollectibles causing you to either lower your margins or raise your rates or both. Payment cycles will change as well, and the company may find itself in the position of having to supply, and pay for, services before they can bill, or collect for them.

If the carrier cannot get credit extended from their underlying carriers, which they probably will not be able to arrange, then they will have to front the money. This creates a gap problem that makes collections all the more important. Given the slim margins involved, and the cost of capital, this can severely impact the financial needs of the company.

Advance Payments

One of the ways that a company can cover itself without going to a true debit system is by an advance payment or deposit system. Both involve a payment deposit for services before they are actually rendered. They differ from a true debit system in that the amount of the advance payment or deposit is not reduced every time a call is made. Instead, the deposit amount is a guarantee against the billing. It is still necessary to monitor the usage, as calls are made, to make sure that the amount of the prepayment, the credit limit, is not exceeded.

Simply stated, in a debit system, a payment is made, and each time a call is made, the amount of the call is deducted from the balance, until it equals zero. In a prepayment system, the amount of each call is added to the total of all the calls made, until the total equals the prepayment amount.

One reason for using an advanced payment or deposit method as opposed to a true debit system is that it's easier to administer, since the real time call rating and credit monitoring systems of a true debit system are not necessary. This form of guarantee is also more acceptable to many corporations. They understand the necessity to give deposits for services to be rendered; they understand the need for advanced payments. Many established carriers also require advanced payments or deposits.

The amount of the deposit is usually determined by the amount of the monthly bill (often an estimate) and the payment cycle. The payment cycle is the amount of time from when a bill is presented to when the carrier receives payment. This can be nearly immediate, if you are applying the bill against a pre-approved credit card or ATM account; or several weeks, if the payment process is rigorous and time-consuming. This overall cycle would add to the deposit required.

For example, if you estimate the amount of usage is to be US$500 per month, then the amount of the advanced payment or deposit would be US$500. If the payment cycle is 2 weeks, this would add another half a month, or $250, in our example, making the total deposit $750. Most carriers will try to negotiate for more, to give themselves a buffer.

Lest you believe that carriers are being foolish and stubborn in this, be advised that losses can be staggering. One US carrier with a very liberal credit policy for resellers, estimated its losses in nonpayments to be over $300 MILLION in a single year. It takes a lot of profit to make up for losses this big. This has caused some odd behavior, however. In one case, a US carrier asked a reseller for a $50,000 deposit BEFORE they would even present a proposal. This could create a unique new revenue stream, if successful. It is doubtful that many companies would be willing to give this type of deposit just to get a proposal.

Deposit money is usually kept in an escrow fund. There are often specific laws dealing with how advance payments and deposits must be treated, how escrow accounts must be set up, by whom they are to be set up and how they are managed. An attorney or accountant should be consulted to make sure that the legal requirements for deposits are met and that the company is operating in accordance with them. If the customer pays the bill on time each month, then the deposit is not touched, and unless the monthly usage increases, the customer continues to pay the bill, and everyone is happy.

However, this all changes if the customer's billing increases enough that the advance payment or deposit no longer covers the monthly bill and billing cycle. In that case, the carrier usually asks for more money to cover the increase.

Another popular collection mechanism is the Irrevocable Letter Of Credit, or ILOC. An ILOC is an instrument that is provided by the subscriber's bank, which is somewhat like a signed check without the date filled in. They are usually for set amounts or maximum amounts and are generally issued for specific periods of time. There is usually a process that is specified in the letter of credit as to what the telecommunications company must do in order to redeem the ILOC. There is usually a seemingly simple mechanism, such as a bill that remains unpaid after 30 days, that allows the holder of the ILOC to cash it in.

Because of the liquidity of the ILOC, the subscriber must make some form of guarantee to the bank that issues the ILOC. This may be anything from securing it with 100% cash to a draw against a line of credit. Large corporations have little problem getting an ILOC and often use them to replace the need for a deposit or advance payment.

The greatest single risk for a telecommunications company is uncollected debt.

Checks, Electronic Funds Transfers, and Merchant Accounts

There are basically three ways for a subscriber to pay the telecommunications company for the services rendered regardless of the guarantee method used. They are checks, electronic funds transfer and merchant accounts.

The use of checks is pretty obvious. They are written either on a foreign or domestic bank, made out for the amount of the bill and sent to the telecommunications company. The telecommunications company deposits it in their account, and after clearing, the money is available to them.

When dealing internationally, there are a couple of caveats. Many countries have restrictions on individuals in that country writing checks to foreign corporations or individuals. It is possible for the subscriber to send you a good check on an account in a good bank with sufficient money in the account to pay the check and send it to you to deposit in your bank only to have it rejected. You then find that the ministry of finance in the country where the check was written does not allow money to be sent out of the country, or the proper paperwork was not done. Many African nations have this kind of restriction. Deal with a bank that has international expertise. They are your best resource for alerting you about potential problems and monetary quirks in various countries.

Another issue that arises when accepting checks from international corporations is the clearing time for those checks. If a check is drawn on a foreign bank, it will probably be payable in the currency of that country. This money is subject to the constant ebbs and flows in the monetary exchange rates. The amount of money on the foreign check may be less than the amount of the US bill.

The other method of sending cash from one country to the other is through the use of some electronic funds transfer, such as an Automated Clearing House (ACH) transfer. For the carrier, the biggest advantage to electronic funds transfer is quick payment. This can also be an advantage to the customer, because it can reduce the payment cycle, lowering the deposit requirement accordingly.

Credit card merchant accounts are the next category of financial transactions, and the most common by far in callback, at least for small business or consumer accounts. This is simply taking a VISA, MasterCard, American Express, Diner's Club, or other card to pay for telecommunications expenditures. There are some limitations to this type of arrangement, which may limit their use to small accounts.

The biggest downside to credit cards is fraud, which represents a risk to carrier and customer alike. If you have an established relationship with the carrier, giving them a credit card may be safe enough, but giving a new or strange company your credit card can lead to many problems later on. For the carrier, the incidence of bad credit cards is enough to make them usable only under the most controlled circumstances.

Billing Requirements

The billing system is the heart of any telecommunications system. The ability to process calls and provide communications is not worth anything, unless you can bill and collect for the calls that are made. Which is probably the most complex function that has to be provided by the system. The communications functions and the customer databases are large, but not particularly complex. The billing function is large and complex! It just seems deceptively simple, which has led to many failed attempts at homebrew billing solutions.

For example: the rate structure is set up to provide a different rate table for each country (there are about 240 different billing jurisdictions in the world). If you establish a different rate table for each country that details the rate to every other country, the total number of routes is 240 X 239, or 57,360 rates. Each of these could have three time periods (standard, discount and economy) and an initial period and subsequent period, for a total of six different rates for each route. Thus, it's possible to have a total of 57,360 X 6 or 344,160 different combinations.

In reality today, with the great increase in special services billing, like cellular telephones, there are over 1,000 rate steps, and growing rapidly. This means that there are over 1,000,000 possible combinations of originations and destinations. This can be even more complex if different calling plans, or discount rates are offered to different customers. This is not a hypothetical example; it's the reality of a competitive telecommunications system today.

There are several approaches to handling this situation, depending upon the exact hardware and systems configuration. One thing to be aware of is the impact on performance that large and complex databases have. A database of the size and complexity we are discussing requires significant processor power to administer. If using a "center stage" switch for the system, you might find that the rating and billing processes are draining the central processor, and causing the entire system to slow down significantly. If you opt for the more complex rating plan, you may have to distribute the function and move it off the central stage platform because of the impact on performance.

Chapter 31
Applications Sell, Technology Doesn't

Don't agree with the title of this chapter? Read on, you will become a believer.

Let's start with the most visible example of this in the telecom world. It started in 1959, at the New York World's Fair. AT&T demonstrated a working telephone system that added two-way video to the system. Now you could not only talk to people, but you could see them at the same time. AT&T optimistically predicted that this new "Picturephone" system would be in widespread use within a few years.

It would start, they said, with businesses, who could better afford this new technology, and then to residences as the technology became more affordable, and people began to see the "benefits" of this technology. I cannot remember a new technology getting the exposure and "buzz" that this did. Everyone was talking about it. There were jokes about showers and dropped towels, and bad hair days, and so on.

What! You don't agree with me? Then give me a call on your Picturephone, OK? What, you don't have one?!? Why not? Doesn't EVERYBODY have one?

Well, the answer is obviously no, not even now when the price of the instrument is less than a meal in a 3 star restaurant, and the transport is free over the Internet. Why not? At Bell Labs, there used to be a standard answer to the question of "why did you invent picturephone?" The answer was, "Because we could!"

AT&T and others introduced, and re-introduced picturephone services for years. In the 1990s, just before I left AT&T, a major sales effort to push Picturephone Meeting Service, was being launched. It flopped, again. Big surprise.

Let's take a look at another technology - Touchtone - an obvious success. In fact, it was a major catalyst in the Bell System moving from antiquated Step by Step central offices to the more modern (in the 70s) #5 Crossbar central office since crossbars enabled touchtone. The extra dollar or two that consumers were willing to pay a month for touchtone actually allowed the Bell System to replace step offices several years sooner than the normal economic life cycle would have allowed.

Bell could have never convinced users to pay for the upgrades just because the technology was better, but give them a new feature (read application) and BAM! They opened their pockets.

Wait a minute! There is a lot more technology to picturephone than touchtone. Picturephone is much "cooler", and represents a new aspect to telephone use, while touchtone is rather boring and is only an enhancement at best. What happened here?

The lesson: no matter how compelling and cool the technology — users make decisions based on *features* not *technology*. The fact is that no matter how mundane touchtone over rotary dial may appear to be, it is far more compelling to consumers than the gee-whiz picturephone.

In fact, let's state the obvious. No matter what the features are, consumers are only going to buy them if they offer benefits. Cell phones are another case in point. It is not the features of cell phones that make consumers buy them, they buy them because of the freedom that they offer.

Let's take a look at a more concrete example. Someone told me the other day, that VoIP, or packet telephone networks would replace legacy PSTN networks in a few years. False! There are billions of dollars invested in today's legacy networks. Even if you concede (and many experts do NOT) that packet networks are inherently better, what does it offer that consumers are willing to pay for? In fact, the biggest sales point is that VoIP is cheaper, so that means that consumers are actually willing to pay *less* for VoIP services than they are willing to pay for legacy services.

What does this say about making investments in VoIP rollout? Why should we spend money to "upgrade" a network, and then have consumers expect to get the services cheaper? This is a legitimate concern, and underlies the weakness we have seen in telecommunications stocks.

What consumers are willing to pay *more* for are applications, driven by VoIP technology. Let's use another telecom analogy. When Electronic Switching System (ESS) central offices became available in the early 70s, consumers were not willing to pay for a better technology, but they were willing to pay, and pay big, for the applications that ESS offered. A good example is call waiting, one of the biggest selling telecom features of all time, just behind touchtone. Call waiting offered a multitude of benefits to consumers, especially those with teenage children, so they bought it in great quantities.

ESS features like call waiting, call forwarding, Caller ID, privacy director (My personal favorite!), speed dialing and many others are found money for the LEC because there is little or no cost associated with providing them, beyond the basic cost of the central office, which they incur anyway.

Applications sell, technology doesn't. Are you a believer yet? If so, read the rest of this chapter, and I will tell you how to make money with VoIP and other next gen telecommunications services. If not, put this book down, and go read *Gone With the Wind*, because that is where your telecom investment will go!

The real power of VoIP is the ability it has to connect to other things, like the Internet. It is the first truly integrated technology, and at long last, finally fulfills the promise of *convergence*. It is this hybrid nature, and the ability to present itself as voice and data, at the same time, that allows the creation of *applications* that consumers will be willing to pay more of their dollars to get.

Let's look at Unified Messaging, which is the Poster Boy for VoIP applications. The concept of unified messaging is that it allows users to retrieve, initiate and handle all messages — whether they are voice, data, application data, faxes, or whatever — from any device anywhere. This is a very ambitious idea, and a very compelling one. Imagine if you could open, create or handle ALL of your messages from one place. Again, AT&T pioneered this effort, most notably in the 80s.

There are some practical considerations to this. For example, how do you retrieve a fax or an email from a telephone? Okay, the system could read the email to you, but what if there is an attachment? How do you see it? What if the message is an image? It will likely take a long time before there is a single, universal device that can handle any message, regardless of the medium used, but in the meantime, unified messaging has some appeal.

Most UM suppliers provide a telephone number for all voice and fax messages to go to. This is a big step in the right direction. Now, add a web interface, and you have even more ability to integrate messages. Also include a wireless device, like a two-way pager or a wireless Palm, and you have a system with some application to the problems of mobile end users.

It is this "layering" of features that gives VoIP the promise that it will be able to stimulate end users. But, I have to tell you, so far the offers and marketing that I've seen leaves a lot to be desired. Most VoIP companies, as with many technological startups, have focused on selling the features and the advanced technology. While this may appeal to a few early users, the hoards of users are waiting for a *raison d'utiliser* for VoIP. So far, all we have given them is engineering data, and reasons why they should pay us for their services. They do not yet see the *value*.

The other lesson we have learned through the opening of telecom markets is that users are imbued with a great measure of inertia, and do not flock immediately to the newest technologies. Take a look at fax machines. They have been around for decades, but the explosion did not start until the mid-80s, when some critical mass was reached. By that time, fax machines were in a significant portion of businesses, and the price had dropped to affordable levels. Today, people do not ask, "do you have a fax machine?" They ask, "what is your fax number?" This attests to the widespread use and assumption that EVERYONE has a fax machine, showing the almost universal acceptance of fax machines. And they are not going away any time soon.

Chapter 32
Quality of Service

"The bitterness of poor quality often lasts longer than the sweetness of low price."

This often quoted, and more often, paraphrased, commentary on value has particular relevance to anyone involved in telecommunications. Service providers struggle to lower their end user prices to meet competition, and users often become obsessed with finding the absolute rock bottom price for whatever service or product they need. It is true that there are applications that do not have a compelling need for superior quality, but the trick is to match the price points necessary to be competitive to reach the level of quality that is needed.

There are several elements of quality that need to be specifically addressed in designing the right service to meet the needs of the application. This is where VoIP, still in its early stages of evolution, offers some significant advantages. As we discussed before, trying to use the Internet as a transmission media can be very troublesome, but it is cheap, and the resultant quality is often sufficient to meet the quality of service standard for many applications. For example, the bandwidth needed to support an application like prepaid calling cards can be very modest. It is not as demanding as fax or data traffic, and can be tolerant of what would otherwise be an unacceptable level of latency, if the price is right.

On the other hand, if you are selling 1+ to business customers with a lot of traffic to South America or the Pacific Rim, you will not be able to get away with less than an excellent quality of service in every way. Why? Because businesses, especially those involved in international commerce, use a lot of faxes, and fax machines are very sensitive to the quality of the communications facility, and a lack of quality can cost money.

Fax machines will step down in speed from 9600 bit per second (or 14,400, in some cases) to as low as 4800, to reduce the error rate. At 9600, each page will take approximately one minute to transmit. Every time the baud rate is halved, it doubles the transmission time. If the user is sending a two page fax, and the fax machine drops back to 4800 bits per second, it will take 4 minutes instead of two. This doubles the cost to the user for sending the same fax. If he had been sold on a 15% savings for the inferior service, he would lose a lot of money on the poor quality.

Mean Time Between Failures

This is usually thought of as a measurement of equipment reliability, but the same principle can be applied to network availability. It is probably the most important technical consideration to users. Telecommunications is a very competitive industry, and since it is price-based, alternative service providers and emerging carriers must strive to offer a value that is at least comparable to what is available in the market. If the alternate service to the PTT is not as reliable, the price advantage will soon be forgotten, and your customers will go to other services or back to the PTT. Even a relatively minor disruption, especially if it occurs at the wrong time, can cost you many customers.

Your switching equipment is going to be enough of a concern in the area of reliability, so don't compound your headaches with a network you can't count on. Even if the carrier's network itself is one hundred percent reliable, you probably have the Local Exchange Carrier in between, and they will cause outages from time to time.

Mean Time to Repair

Probably this is the second most important technical consideration. Of course, if the network never goes down, you don't have to worry about the repair interval. But when you do need repair, whether it is for a crashed network, or the LEC has lost the circuit, or even a translation snafu, you want to know that you can count on it being corrected in a short time frame.

Remember that the carrier that you deal with may have to rely on several other organizations to keep the service up, so you want to make sure that you are very comfortable with the capabilities and motivation of the carrier to respond. In the past, dealing with any telephone company could lead to frustration when non-cooperative attitudes were encountered. The overall quality of the repair function probably is more a matter of motivation and resources than technical knowledge, so look for these traits first.

Size and Deployment of Service Personnel

As we said before, the overall performance of the carrier in the area of repair is highly dependent on the attitude of management and staff and the amount of available resources. A trip to the Point of Presence (POP) and the service center should reveal the attitudes of the company on repair. But you also need to ask a few, in fact, a lot, of questions about how much manpower is available and where it is deployed.

It is of very little comfort if the carrier promises a two hour turn around time on repair, and the lines are always busy. Turn around time is defined as the time between when a user reports a problem, and it is fixed. If it takes the user an hour to report the trouble, you have delayed the start of any meaningful resolution of the problem for that period. This is a common problem. When Hurricane Andrew hit South Florida in 1992. It caused wide disruptions in all services, including telephone. Service centers were jammed for days and even weeks afterward all over the country, not just in Florida.

After such a catastrophic event, those in the wake of the disaster will understand long delays from service centers to respond and in the restoration of service. But the customer two thousand miles away may not be so understanding or sensitive to your situation. Make sure you have asked the difficult questions and that you have gotten commitments, not just statements of what the carrier will "try" to do in the event of an emergency.

Call Setup Time

For callback, call setup time is a critical measurement. This is the time from when the user pushes the last digit to when it begins to ring on the far end. A short setup time is not so important on the first leg — the subscriber leg — as it is on the second or destination leg because the time is running on the subscriber leg while we are waiting for the second call to complete.

Whether you charge for the overhead time of the subscriber leg or you absorb it into your business plan, it drastically affects the cost of a completed call. The time will vary country by country and is related to the quality of the network on the remote end. However, there is a very large and noticeable difference in the set up time from carrier to carrier for a given country. Some carriers, notably AT&T and MCI Worldcom, have consistently shorter set up times than other carriers.

Overhead time can be measured fairly easily in the US by making a series of calls on a POTS line using the Dial Around (1010xxx) Code for the carriers you are considering. Simply dial the equal access code assigned to the carrier, then "011" for an international call, the country code, any city code, and the telephone number for the carriers you are interested in. Begin timing the interval from the time you push the last digit, until you hear the ringing on the far end. In order to do this accurately, you must make a series of calls on each carrier at different times and then average them.

Remember that the time that you hear the ringing on the far end can vary up to six seconds depending on where you fall in the ringing cycle, but if you make a few calls you can minimize this impact.

Some carriers may claim that a nodal service, like T-1 is faster, since the time to access the carrier through the LEC is included when you dial the equal access code from a POTS line. There may be some validity to this claim, but if the carrier has a significant delay coming through the LEC, it is likely that there will be other bottlenecks in the network as well.

International Call Supervision

Call supervision is the process of controlling a telephone call. For example, if party A calls party B, when party A hangs up the call, the circuit will disconnect. Without call supervision, you would only be able to make one telephone call when you got a new telephone line, then you would not be able to hang up. The other side of the coin, when the called party, or party B, hangs up, we want to notify the central office that party A is still connected to the disconnected called party. Sometimes this is referred to as "answer supervision", since it indicates when the called party answers or disconnects from the call.

In callback, this feature and the way it operates, takes on added significance for two reasons.

First, the timing of the calls is critical. It is vital that when one party or the other hangs up, the call is disconnected in a TIMELY manner. "Timely" in this context means no more than a few seconds. If you have a carrier that bills in 1/10 minute, or 6 second increments, one added second could cost you 1/10 of a minute, which could be as much as fifty cents, or more. Add to this the time for a call that doesn't complete, for example, a busy number, and this could cause your expenses to go out of sight.

The second reason that this is vitally important especially in callback, is that both legs of the call are outward from the service provider, so even the user is the called, or "B" party. This means that if the user, your customer, doesn't properly disconnect, or just hangs up the telephone, the billing timer could run for several more seconds.

What is critical is the quality of call supervision, and for international calls, it is difficult. Without going into the details, variations in the networks amongst the countries make it a very difficult process to assure. It requires all of the diligence and technical expertise a carrier can muster. For this reason, the quality of call supervision is carrier dependent, and varies greatly between carriers. There have been instances of carriers withholding answer supervision for up to a minute and a half after the called party disconnects, so you must be very careful to not only get the carrier's commitment on answer supervision, but to measure and monitor it frequently.

Make sure you know the level of quality in the call supervision of the carrier you are dealing with, and again, get their commitment in writing, if possible.

Intelligent Front End Maintenance

In the current competitive environment, the carriers face a dilemma: on the one hand there's the need to cut costs, but on the other hand, there's the need to provide a high level of customer service. Different carriers have enacted different strategies, but the bottom line is similar. Provide the highest level of service to the biggest and most profitable customers, while assuring that no group of users have any concerns about the quality of the service provided.

For this reason, many carriers have built very intelligent systems to front end their service business. This ranges from the Interactive Voice Response (IVR) systems that many carriers and LEC's use to take service problems automatically, to a more proactive, network management oriented process.

Automating the process of receiving and registering service complaints is one thing, but actually striving to keep the network operating is another.

In the first instance, we put the responsibility for detecting and fault isolating a problem squarely on the end user, making responsiveness to situations on the part of a carrier a reactive posture. If the end user agrees to take on the burden of managing the process, the carrier saves a lot of time and customer frustration. This is the way most small to medium size business are handled. The provider of the service installs the service, checks it out, and turns it over to the end user.

If there is a problem, the end user ascertains whether the problem is in the equipment, the network, the power supply, the system software, or where ever else it may be. The user then calls the vendor they think is the source of the outage and reports a problem. The user gets a commitment from the vendor as to when they can expect the problem to be resolved. It is then up to the end user to track the commitment and make sure it is resolved. If the problem turns out to not be the element of that particular vendor, the end user must then call the next suspect and begin the process again.

An alternative to this cumbersome and labor dependent process is a service assurance program provided by the supplier. In the case of a network vendor, such as a carrier providing T-1 facilities to a callback company, this represents a commitment on the part of the carrier to up time, and not just a repair response.

How does this work?

In the case of nodal network facilities, such as T-1's provided for reseller use, the network vendor can just as easily detect problems with the service as the end user can. This is because the T-1 is a point-to-point service linking the carrier to the company, so either end can detect a fault. In fact, with T-1's, loss of signal shows up at both ends of the circuit at the same time. The equipment in the carrier's POP is normally arranged with some form of automated network management system that will detect and attempt to automatically correct any network faults. If it cannot fix the problem within a given interval, it will automatically escalate the problem, to a human being, if necessary.

This type of service can take a big burden off your management team, since the network provider manages the process. Even more sophisticated network management systems can detect problems, such as a sudden change in the traffic load, or the sudden loss of any traffic at all. These advanced features can even help with network security and prevent hacker attacks, etc.

When talking with carriers about their service, be sure to ask about these advanced features. There may be some charge for them, or they may come with a certain level of revenue or usage commitment. The fact that a carrier offers them at all should give you some insight into the carrier's perspective on customer service.

Diverse Routing

In order to provide an adequately robust service, most carriers strive for diversity in their routes. Diversity means that no place in the physical network has only one path out. This prevents what is known in the industry as "backhoe fade". This refers to the fact that one backhoe on a construction site can disrupt the network by accidentally breaking the cable in the ground.

There are many facilities-based carriers with routes that have only one way in and out of a geographic region. If you are behind a situation like that, you may find that one errant backhoe can disrupt cable for several hours or more. Even switch-based carriers can have this problem. Several carriers were out of service for days and even weeks in early 1994 when flooding hit the Midwestern and Southern regions of the US. An outage of this duration would lose most international resellers a significant portion of their customer base in a very short period of time.

While no carrier can guarantee full diversity on every route, be sure you know where the carrier stands in the location you are interested in.

Traffic Handling Capability

While it may not be obvious to people who are not involved in telecommunications, there are limits to the traffic handling capabilities of every carrier. No one, including AT&T, is immune to occasional blockages. If you don't believe it, try a call on Christmas Day or Mother's Day in the US — you can be assured that every network has periodic overloads. When the stock market took a nosedive in October 1987, most of the US telephone carriers, particularly in the Northeast US, had extensive blockages for hours.

While most carriers have sufficient capacity to deal with periodic bursts of traffic, there is a limit to what they can handle without disruption. It is important to know what the carrier's record is regarding overloads. Even more important is knowing what the carrier's capacity is on specific routes, particularly if the underlying carrier is not one of the "big 3", AT&T, MCI Worldcom, or Sprint.

A switch-based reseller has to buy circuits and/or bandwidth into and out of specific countries and routes. They often buy extra capacity in terms of backup to other carriers, but there are definite limits. The traffic that a new international long distance reseller brings to the carrier will be a noticeable load, but whether it is enough to adversely affect traffic to that area needs to be discussed with the carrier. If your projections are complete and detailed, the carrier should be able to determine if the traffic will have a measurable impact or not.

Type of Facilities Provided

There is a major issue in international long distance reselling, particularly call-back, which must be addressed and understood. It is the issue of echo, or delay. In the early days of long distance, many carriers, particularly AT&T, began the deployment of satellites. It was widely believed that satellites were the ultimate transport medium for global communications. They were not aware of the impact of one vital element of satellite communications.

Geosynchronous communications satellites orbit the earth at 22,300 miles in space. Since the signals from the earth have to travel up to, and back from the satellite, they must go a total of 44,600 miles. These signals travel at the speed of light, 186,000 miles per second, which is clearly very fast! In fact, scientists are still not sure if it is possible for anything to go faster. But as fast as that is, it still represents a time delay of over a quarter of a second going up and coming back. The consumer recognized long ago that this seemingly minor delay was somewhat annoying.

In fact, it became almost intolerable. That is why Sprint, AT&T, MCI Worldcom and others spend a great deal of money convincing the public that fiber optics is better. In fact, it is, at least where delay is concerned. If we put a fiber cable around the earth, it would stretch a total of 24,887 miles. But, even if a signal was sent via this cable for the whole 24,887 miles (which probably would never happen), it would still only present to the caller a delay of one eighth of a second. At most, it would go half that or 12,444 miles, which is a delay of .0665 seconds — not enough to worry about.

The complication that callback and international long distance resell brings to this is that each call is composed of two legs, so if there are satellites involved on both legs, the delay could be over a half a second. While this may not seem like a lot, an ITU study in 1992 suggested that a 350 ms. delay, or about a third of a second is enough to significantly disrupt voice communications. This would tend to argue for carriers that use fiber extensively, and avoid the use of satellites.

All carriers have to use satellites to some extent, since fiber is not available everywhere. Of course, the big three in the US have access to more fiber miles than other carriers. If you are working with a carrier that obtains network facilities from others, get a feel for how much fiber versus satellite you are going to face, and try to minimize the use of satellite.

As you can see, many of these items are difficult to measure; a more qualitative approach is needed. Asking the sales person to describe their capabilities in the technical arena is not likely to yield very reliable data, since they are trained to sell the service despite objections. It is not to say that the sales person would intentionally mislead a customer on the technical quality of their offering, but it is likely that they really do not know what the differences in their offering and those of their competitors actually are.

Terms and Conditions

Almost as important as the technical capabilities of the carrier are the terms and conditions of their offerings. Depending on the carrier, and how bad they want your traffic, the terms and conditions can be even more difficult to negotiate than the price. The flexibility of individual carriers will vary from the AT&T "its in the tariff" stance on many issues, to a "whatever it takes" attitude on the part of newer or hungrier carriers.

Don't be fooled into thinking that the willingness of a carrier to be flexible is related to the quality of service they intend to provide. Remember that in most cases, you can only see into the soul of the carrier through the eyes of the account team you are dealing with. If the request for flexibility is occurring while you are in a buying cycle, you will likely find them much more flexible before the sale is made than afterwards.

The point is, be sure that you ask for whatever you feel you need before you commit to any agreement with any carrier. While this may sound like rather basic advice, be assured that many enterprises have found themselves in the position of begging a carrier for a small consideration that they could have had by simply asking for it at the right time. And the right time is before you have made the commitment.

Be sure to review all the relevant documents that will define the relationship. If the carrier is proposing a service that is part of a tariff, ask them for copies of the tariff. Many people have gotten themselves into a bind because they didn't think that the tariff applied to them, or because they thought the carrier could waive portions of the tariff. A tariff is a contract, even if no one asked you to sign it. It is applicable to all, even if they were not aware of it; and enforceable, even if they did not specifically agree to it, or were not told about it.

AT&T is currently the only carrier in the US that must get tariffs approved in advance with the FCC. The others must file tariffs, but they are advisory tariffs because they can be acted upon pending filing. It is also true that the detail of the AT&T tariffs is much finer because of the history of regulation. Every detail of the relationship and the offering is spelled out giving a complete picture of the offering. There are, of course, times when AT&T has the flexibility to deviate from the written specifications, but these are rare. One of the factors, which make it so rare, is that it is easier for the AT&T people to adhere to the tariff than it is to make a decision to change anything.

In the meantime, having a rigid tariff is a mixed blessing. It binds AT&T to the wording of the attorneys and regulators. This means that there can be very little question as to what the structure of the offer really is, but it also prevents the carrier from showing the flexibility that may be needed by a fledgling call-back international long distance reseller. As has been emphasized, the key to success in the market of the future is to have a high level of flexibility in all phases of the relationship.

If your agreement is not part of a general or custom tariff offering, make sure that you spell out everything that you expect and need from the vendor. This includes the manner and timing of billing, how disputes are resolved, when and how payments must be made, specific remedies in the event of service interruptions, levels of service that can be expected, and so on.

Customer Support

There are many aspects to the amount and type of support that is needed by an international long distance reseller. It is vital for this type of reseller to have a generous amount of vendor support available to them. It takes many forms, and addresses many needs. You will want to ascertain the vendor's level of support and relative competence in all of the different areas that they have to support. Carriers have the resources and depth of knowledge to assist in avoiding and rectifying many situations. It may be difficult to quantify the relative competence of the vendor, but you must be thoroughly familiar with the vendor's support. More important, you must have confidence in the vendor's ability to provide the level of support you need.

There is a need for pre-sale support to design and recommend the right type of service. Most vendors are very good at providing this type of support, and can advise you on the best service arrangements for your business. Because of the changing nature of the international long distance resell business, it would behoove you to make sure that you have access to these resources on an on-going basis. Many vendors will be your best friend at the time that you originally buy their service, but fade away later. Don't be afraid to stay in touch with the account team, and get up to date on rumors, ideas and, yes, gossip.

Once the level of the service is agreed upon, the carrier must do a superior job of provisioning the service and assuring that it is functioning properly. This is frankly the weakest link with most carriers. If you don't feel that you have the experience or knowledge to do your own implementation, and correct any problems, you should make sure that you have confidence in the carrier to do it for you. You have the right to expect that they will successfully install and assure the service. Ask to meet the project manager or managers, and determine for yourself whether they can get this network project working in your behalf.

There is also a need for a high level of customer service in billing and collecting. Not the bills you are sending to your customers, but the bills you will receive from the carrier. It has been stated that the biggest issue in long distance is billing. This has probably always been the case. There is no system that is more complex or demanding as a long distance billing system. The many different routes, times of day, multiple accounts, discount packages, different time intervals, and so forth make long distance billing reconciliation a very tedious process.

It has been said that the vast majority of the long distance bills that are rendered have some error in them. International long distance resellers have the dual responsibility of reconciling the bills that they send to their customers with the one that they receive from the carrier. You have to rely on the assistance of the carrier to help in clearing up your bill, because you will find a variety of problems, including calls that you can't account for, inaccurate rates and the dreaded "stuck call". This is a call for which the disconnect supervision didn't function, and the carrier is billing you for a call to China that is several days long. These problems can be very time consuming to reconcile and adjust, so you should try to arrange to have a contact in the customer support department at the carrier to assist you in getting the invoice correct. Again, make sure you understand and can accept the level of assistance you will get in this area.

And finally, there is a need to provide a certain level of customer service in the event of a service problem or disruption. Since you are repackaging and selling the service to your end users, when you have a service problem, whether it is service affecting or just annoying, you will want to have a high level of confidence in the ability of the carrier to locate and clear the problem. Ask the carrier to specifically enumerate the steps they will take, including the escalation procedures, and actual performance data. Objectives are no good here, since they only represent the promise made, not the promise fulfilled.

Also beware of the term, "response time" as opposed to "repair time" or "turnaround time". Response time could mean that you get a call from a test technician who is working on the problem. Working on the problem does not mean that service is available again, just that it has been put into the works to be repaired. Ask the carrier to provide statistics on actual repair times in situations that are analogous to your environment.

Chapter 33
Free Long Distance

This is truly one of my favorites. Granted that until recently long distance has been irrationally high since the invention of the telephone. In many spots around the world it is still high, due to anti-competitive regulation and artificial market manipulation, but left to its own devices, it drops drastically. Couple this with the advances in network and fiber optic technology, which have dramatically lowered the costs of providing service, and *vióla!* the price goes into a free fall — that's what is happening now.

But free? Come on! Name one commodity that dropped in price, and was ultimately given away for absolutely free? I can't think of a one, and neither can you. It just doesn't happen. There is even a price for the salt and pepper in a restaurant. As miniscule as it might be, if every patron quit using it, the price of a meal would drop, given a reasonably competitive environment. Nothing of any value is free.

Can't happen, won't happen. Can you visualize what the television advertisements would look like? "With Hunky-Dory Long Distance, You Can Count on Us Because We're Free!" Take a look at cellular telephones. There are a lot of ads now in the US and other places where the set itself is $1.00, or "free." I saw one that said the cellular company would give the buyer a dollar to take the phone. I was going to buy several million of them, until I discovered that you had to sign up for the cellular service at $XX per month, EACH. Nothing is ever free, if it has any value at all.

There really is no precedent in telecommunications of anything becoming free. Local calls and the facilities required to make them have, in fact, increased many times in price, even since the advent of "competition." (Even though a lot of attention has been given to increasing competition in local communication markets, i.e. CLECs, in reality there is very little competition in the local markets anywhere in the world today.) What does happen is that new products and new services are added in order to fill the revenue gaps.

For example, take a look at your local telephone bill today, the line charge is probably not a whole lot different or maybe just a little more than it was 15-20 years ago. What has increased is the number of auxiliary charges that are now added to the bill for "necessary" services such as caller ID, not being listed in a telephone directory, voice mail, call-waiting and other services, which are deemed to be essential today. What also happened with the advent of the Internet, many homes now have two telephone lines, one for voice calls and one for the Internet. Nothing has been provided for free. In fact, a telephone bill today in 2001 will likely be three to four times what it was 15 years ago because of all the extra services, which have been put in and bundled.

Instead of becoming "free", some long distance telecommunications services may move to different pricing models, i.e. flat rate pricing or long distance bundled with other services. The siren song of free voice telephone services over the Internet will not become a reality anytime in the near future, either. This is because the bigger the Internet becomes, the more bandwidth hungry services, i.e. broadcasting, audio, video, etc. will move in and gobble up that bandwidth. To assume that there will be enough reliable bandwidth to carry even a faction of the world's voice traffic is an unrealistic assumption. There are too many bottlenecks within the system and too many places where the carriers have no incentive for adding bandwidth because of the pricing structures that are in place today.

It is likely that in the next few years, we will continue doing the long distance rate limbo (how low can it go, now?) and every time it goes down another order of magnitude, somebody will say it can't possibly go down any further. But it will. And when it does, someone will boldly predict that long distance will become free, and be widely quoted, but it will turn out that it will have been an incorrect prediction once again.

Chapter 34
ENUM Services

The final convergence of telecommunications and the Internet may be ENUM services. This is a new initiative started by the same fine folks that brought us the Domain Name Service that humanizes the Internet. But it actually started in 1993 within the confines of the Internet Engineering Task Force (IETF). ENUM is the name that was adopted by the telephone numbering working group of the IETF to describe a mechanism using the Internet Domain Name System (DNS) to map E.164 (an ITU standard that describes the format of telephone numbers worldwide) numbers to URLs. The basic idea was to create a way to resolve telephone numbers over the Internet.

This concept may appear to be a bit esoteric, as with many of the concepts that underlie the Internet. The practical applications of ENUM are very appealing, especially to VoIP. The impact on Competitive Local Exchange Carriers (CLECs) could be very important.

In order to understand ENUM and its applications let's look at a couple of the considerations involved. First, the DNS system is used in the Internet. Sites on the Internet are identified by a series of 4 hexadecimal numbers. For example, the IP network address of the server for www.microsoft.com is 207.46.197.100. The IETF decided early on to create aliases for these numbers because human beings are more oriented to names than to binary digits. In fact, the IP address, 207.46.197.100 is really a representation of a real address

— 11001111001011101100010101100100. I don't know about you, but the binary number is almost impossible to remember, or type in consistently. But it is what the binary world understands. 207.46.197.100, is the hexadecimal version of the binary number, a bit easier for human beings to handle, but still not very memorable.

To make this easier and to expedite communications, the Internet uses DNS servers, strategically located, to allow the use of names like www.microsoft.com to be used to address web sites. When you type one of these Uniform Resource Locators (URLs) into an Internet browser, it goes to the nearest DNS server, retrieves the binary address and uses that to get to the actual web site for Microsoft. (There are actually a couple of other steps involved here, but we will leave that to Internet geeks to explain.)

There is a second important advantage to the DNS system. It allows web sites to be moved to different servers, without changing the URL. If Microsoft moved its site to a different server, it could still use www.microsoft.com as its web site address (URL), but the network DNS servers would *resolve* this to a different binary address. It gives portability to Internet sites.

OK, you remember *portability* in a different context? Telephone number portability was a major objective of the 1996 Telecommunications Act. After its passage, the FCC created a National Number Portability system, to allow telephone users to take their telephone numbers with them, wherever they might go in the country. It was an important initiative because surveys have shown that the vast majority of users will not change their local service to CLECs if they have to change their telephone numbers in the process.

The initial concept behind ENUM is to define a DNS-based architecture and protocols for mapping a telephone number to a set of attributes (e.g. URLs) which can be used to contact a resource associated with that number. Telephone numbers are one long number sequence that human beings have been conditioned to use and remember, up to 14 decimal numbers. (Some of us are old enough to remember when telephone numbers were names plus numbers. WAlnut 0825, was the first telephone number I can remember.)

There are billions of touchtone pads in use around the world. They are the most ubiquitous type of communication terminals in existence, but they cannot be used to access the Internet. Nor can the Internet be used to access most of these telephones. This "Ion Curtain" between the PSTN and the Internet, is a common problem, and inhibits a symbiotic relationship between the two that could exponentially grow them both. Enter ENUM, which promises to bridge the gap, building on the strengths of each one. By enabling the use of traditional telephone numbers on the Internet, it will remove some of the user resistance to making the change.

How does this enabling occur? First, ENUM formats the telephone number in a form that is understandable by the NAPTR, Naming Authority PoinTeR, a fancy name for an Internet database that stores information about the destination number, including its capabilities — H.323, SIP, VoIP, fax, email, web sites, paging, and so forth. It then redirects the ENUM request to whatever is needed to access the appropriate resource.

Maybe an example would help!

Let's say that the telephone number 700-555-1212 is a pager. You are on a web browser and want to page this person. You would enter a yet-to-be-determined address into a browser, and the network would resolve this to a server with access to this pager. Depending on the application, you might have to enter your callback number, or the message. Much of this is not decided today. Bottom line is, you can page to and from the Internet and between the Internet and the PSTN.

This is truly convergence, and ENUM is probably the most serious approach so far to implementing the convergence of the Internet and the PSTN. It could be a major opportunity, if as some pundits claim, ENUM becomes the next "killer application" to tickle the fancy of millions of avid consumers.

Chapter 35
How the Heck Do You Make Money?

The number one concern of every telecommunications company is, or should be, *how the heck do you make money?* Unfortunately, before the telecommunications stock crash of 2000, many companies did not focus on this basic business need. There was too much easy investment money available, so many companies forgot about making money, and tapped investment money instead.

This led to a "carpe diem" attitude on the part of telecoms, big and small, who began to believe that no matter what they did in running the business, they could always get more money. And this was true for over five years. Whether the result was the fault of spoiled telecommunications companies, or naïve investors, is irrelevant, since success in the future, both in running a company and in raising investment money, will be highly dependent on showing a good track record of profitability.

So, how the heck do you make money in telecommunications anyway?

That is the important question. The answer is not simple, but a few things are clear. First, you DO NOT make money by:

* Losing money on each minute sold, but "making it up in volume."
* Pursuing an aggressive growth program with no notion of the size of the opportunity, or where it is.
* Attempting to run a competitive telecommunications company like a regulated monopoly.

203

* Buying private jets, equipping offices with Jacuzzi's, having employee cruises or sponsoring a Beatles concert at the next industry conference.
* Forgetting to look at the bottom line.
* Ignoring the need for investors and stockholders to see a decent return on their investment.

There are probably a few hundred other equally dangerous things you can do to assure that your company will not make a profit, but I do not want to waste time focusing on them. Let's turn instead to the values that you want to maintain in order to keep your eye on the *profit* ball. These are general and apply to any business, but especially in the over funded, glitzy world of high tech, and they are worth repeating.

* **Have a business plan.** Not one of those fluffy, "two pager", buzz word ridden mantras that you see all the time. Have a real, detailed, living plan of what you are going to do, how you are going to get it done, when it will be done and the specific measurements that will indicate if you are on track or not. It also needs a short term, i.e. one year plan, as well as a long term focus.

* **Make sure you live the plan.** Writing a tome the size of *Encyclopedia Britannica* and parking it on the CEOs desk may be impressive to an anal retentive Ph.D., but will not further the cause of the business an iota. You must communicate the plan, assess your progress *daily* and update the plan as conditions change.

* **Motivate and lead.** Even a one person business has at least one employee, who deserves to be treated as the company's most important resource. If you ever think your most important asset is anything other than the employees, it is time to get out. Your employees can make or break your company. Make sure that they are motivated, clearly understand their role and mission, and are equipped to achieve their objectives.

* **Know thy customers.** If this is not the 11th Commandment, a saying of Confucius, or a Gita, it should be. You must know your customers, who they are, how they make buying decisions, why they buy, where they are, when they make decisions, what is important to them, and so on, and so on. You can never know too much about them. Focus your products on what they want and need.

* **Know thy competitors.** This is second only to knowing your customers in importance. You should be as versed as they are on their offerings, how

they function, what they charge, who they are and where they go. Any military general will tell you that if you forget to consider your enemy, they win. Study them thoroughly, and go see the movie, *Patton*, again. It is a great lesson in competitive strategy, and maybe, leadership.

* **Those who live by the numbers, die by the numbers. Those who forget about the numbers, die.** You cannot manage only the numbers, but you cannot manage without the numbers. This Yin - Yang approach to the metrics in any business may sound a bit like "Zen and the Art of Managing a Telecom Company", but it represents the balanced approach to the measurement systems in a business that is necessary. You must study the numbers to see what is happening, but you also must manage the people to affect the numbers. If you try to manage the numbers, you will find that they do not respond. The trick is to identify the meaningful numbers, and then to take the actions that are necessary in order to influence them in the direction you want them to go.

Lest you think that these principles are old fashioned and superfluous, review them in light of the high tech bottoming out that occurred in 2000. Which of the failures followed these fundamental principles? Which of the survivors did not? In telecom, the answer to both questions is "none." In the past, telecom companies could survive either through pure growth, constant rounds of equity investment, or a combination of both, since one feeds off the other. If operations failed to yield enough cash to fund itself, no problem, just sell some stock, or issue some bonds. From 1995 to 2000, when investment capital was plentiful, many startup companies adopted this approach.

But, even in the face of a glut of investment money, this "promise and spend" strategy was doomed to failure at some point in time, if a significant revenue flow and decent margins did not result. A company cannot funds its operations off its press releases forever. The investment boom that endured through the latter part of the 1990s did allow a number of companies to put off what would have, and should have, been a quick end. Instead, they were able to grow and expand by attracting investors. One well-known company had revenues of $1.3 billion over a 5 year period from 1995 to 2000, but expenses of nearly $3 billion over the same period. Clearly not a formula for success.

Towards the end of the year 2000, the bubble burst. Unlimited investment capital sources dried up, either depleted or reluctant to continue making unrequited investments. Telecommunications started to develop a reputation for big losses and incredible risks. Since the stock market runs on rumor and suspicion, the negative "buzz" that began to cloud the high tech investment skies spilled over into telecommunications. "The bigger they are, the harder they fall." Telecommunications, being a huge and important sector, began to fall, and fall hard. This telecom malady spread like a cancer to the rest of the stock market, especially high tech stocks, and the rest is history.

By the dawn of 2001, fueled by glum revenue and earnings projections being made by the largest and most influential telecommunications companies, telecommunications stocks started a free fall. This crash ended up with some small startups losing over 99% of their value, and even affected industry leaders, like AT&T, which lost over 60% of its value in less than a year. Northern Telecom, Nortel Networks, dropped from the mid-80s to less than $7 in one year, a loss of 92% of its value. Nortel had risen, like a rising star from the 10-20 range in less than a year, held it there for a short time, then dropped rapidly.

And bankruptcies. Unbelievable. In 2001, there was a spate of failures that boggled the mind. Viatel, Ursus, World Access (which included World eXchange, Facilicom and others that it had acquired) and Star (after a failed merger attempt with World Access) were highly visible, large, publicly traded companies that went under. They represented billions of dollars of investments. There were scores of privately held companies as well, too numerous to mention, but one Tier 1 US carrier has been rumored to have lost over $400 million in reseller bad debt in 2000.

All of this caused investors to refocus their interest in telecommunications on the basics: revenue growth and profitability. Just as the shine faded from buzz word laden Internet business plans, savvy investors have begun scrutinizing telecommunications business plans, looking for solid indications of profitability. Ambition is admirable, investors want companies to strive for success, but the business plan has to focus on solid, attainable profit-driven factors.

So, to get back to the question at hand,

How the heck do you make money anyway?

What are the elements of profitability in telecom?

Is there a magic formula for achieving profitability in telecom?

How can one identify telecom companies that are likely to make profits and succeed?

In order to answer the issue of profitability in telecommunications, let's take a look at some of the more significant segments of telecom, and make some general comments about their profitability.

Local Exchange Carrier

This is the Valhalla of telecom, being the only sector that is still relatively unscathed by competition worldwide. I am sure that the incumbent LECs might disagree with this sentiment, but even in the US (which is one of the countries with the highest penetration of local competition), as of December 2000, CLECs account for only 8.5% of the local service lines, nationwide. (*Trends in Telephone Competition*, Common Carrier Bureau, August, 2001) This is after nearly 20 years of formal competition, and 5 years of "...incrementally powerful vehicles for competitors to enter local service markets...", a reference to the 1996 Telecommunications Act.

In this same report, CLEC revenues are estimated at 5.8% of the total market, indicating that CLEC have to discount heavily to compete. The 2001 fallout of CLECs, which dropped into bankruptcy like flies, indicates that, under the current structure they are unable to survive. This is even more interesting in light of the fact that the ILECs (Incumbent LECs) are fighting hard to get into the long distance game here in the US.

The 1996 Act required that they be able to demonstrate that the local service markets were open to competition *before* they were permitted to offer in-region inter-LATA long distance service. (This means calls that are outside one of the 161 Local Access and Transport Areas, but within the region.) Looking at what happened, it is difficult to understand how some jurisdictions allowed their ILECs to offer long distance. But the fact is that five were approved by the FCC, including one (Oklahoma) that had a 6% penetration of CLECs, before the fall.

The driving factor here is the cost of the infrastructure for providing service to the so-called "last mile", the facility that actually delivers service to your home or office. The ILECs use the wires or fibers that they have built, using their monopoly revenues, to deliver these services. There is also the matter of central offices that have to be built and maintained.

ILECs today get into the business in two primary ways. First, they lease the last mile facilities from the LEC. This had been the model of the startups, but, alas, these startups have largely failed to survive. The other way is for well-funded carriers, often long distance carriers or LEC from other places, to build fiber rings in highly dense locations like big cities, and offer local service over these facilities. The second model has succeeded.

Clearly, the profitability lesson is that while ILECs are very profitable, most CLECs are not sufficiently profitable to survive. This is largely due to the fact that the ILECs still have a virtual monopoly in the services they offer, even with some apparently aggressive regulatory actions. Just look at the US — despite concerted efforts by the FCC after the 1996 Telecommunications Act, which mandated local competition — it's five years later and the ILECs have lost only a minor fraction of their market.

There are obviously other significant barriers to entry, so ILECs should remain profitable, as long as they can maintain these *de facto* monopolies, even where competition is officially allowed. Still, since it is a large and profitable market, it will attract motivated entrepreneurs and clever technologists. There is a fertile ground for new approaches to LEC competition.

The local services profit model is a capital intensive startup, followed by profitability, if the maintenance costs stay in line, and prices do not erode. Even powerful LECs and PTTs could be at risk, if unfettered competition starts to erode price support.

Watch for the advent of viable and vigorous local competition as being the next major battleground in telecommunications.

Wireless, anyone?

Cellular Provider

Cellular or mobile services are unique in telecommunications in that most countries allowed competition almost from the beginning. In the US, for example, half the mobile frequencies were allocated to non-wireline carriers. In the UK, the dominant mobile carriers (BT and Vodafone) are required to offer wholesale minutes to resellers.

Since the advent of mobile services, there has been more competition than with CLEC services. But because there is still a limited resource in the frequencies involved, market entry is still restricted by regulatory fiat. Following our principles, where there are regulatory barriers to entry, the level of competition is constrained. Where competition is held back, prices stay high. Mobile prices around the world have fallen, but not as precipitously as other sectors where the competition was more intense.

At the risk of sounding like a broken record (remember them?), if prices are artificially supported, profits will be robust. Thus it is with the cellular mobile service providers, they have made robust profits where they offer mainstream services. The purveyors of other wireless services, so far, have not met with the same success.

Especially in Europe and the US (although there are other locations), the regulatory authorities have attempted to derive revenues from the sale of mobile bandwidth. This has normally taken the form of an auction, where the frequencies are auctioned to the highest bidder. In the US, this was an unmitigated disaster. The prices bid were so high, that for the most part, the successful bidder has either failed to raise the capital needed, or gone bankrupt in the attempt. In Europe, so-called third generation, or "3G" licenses have been offered. It remains to be seen what the outcome will be, but suffice it to say that the prices paid have been comparable to the US prices.

Now, the intriguing notion is that wireless "last mile" services, or "wireless local loops", could be an answer to the LEC monopoly dilemma. In some countries, where building wire or fiber landlines would be prohibitively expensive, wireless local loops are used for primary LEC services, proving its utility for local service provisioning.

The wireless profit model is very similar to the local service model, high initial investment in physical plant, followed by years of maintenance oriented costs. In mobile services, the potential for offering other services with low incremental costs exists. The individual margins on cellular enhancements is enormous, since the actual incremental cost of providing the enhanced service is miniscule compared to the charges to the end user. One US cellular operator charges $2.95 per month for delivering Caller ID, a feature that probably costs more to block than to actually deliver.

International long distance calls have an enormous profit potential for wireless operators. Take a look at the cost of a one minute call to the Philippines.

Commodity Exchange Price	.103
Tier 1 Carrier Wholesale Rate	.1350
Tier 1 Carrier Retail Rate	.2600
VoIP Carrier Wholesale Rate	.1014
Discount Carrier Retail Rate	.2200
Prepaid Calling Card Retail Rate	.3800 *
Cellular Telephone Retail Rate	1.23

* At first blush, the prepaid calling card rate may look high, given that you see offers like "92 minutes to the Philippines for $10!". Unfortunately, the hidden charges quickly erode this, so we have used a very popular discount prepaid calling card, with no surcharges, a one minute minimum, and no other hanky-panky.

Figure 34. Comparison of Long Distance Charges.

The cellular operator is clearly getting a significant margin on the call, more than four times the next highest retail rate, and over ten times the probable cost of the call.

Basic cellular is highly profitable, add-ons offer limitless opportunities for incremental revenues, and the field is limited by regulatory limitations.

A clear winner!

Long Distance Carrier

Until a few years ago, this was the most profitable service offered by telecommunications companies. The largest telecommunications companies, Tier 1 carriers, and the PTTs around the world, had a lock on a very profitable market. But, callback and VoIP have changed all that, so the international carriers around the world have seen what competition means, up close and personal. And the effect has been dramatic. The failure to respond to these changes and generate profitable revenues have plagued these newly competitive carriers for the last several years, leading to the demise of many of them.

Of all the profitability events in telecom, the change in status of retail long distance from cash cow to commodity competition, ranks as the most drastic change in the past decade. This event is probably the biggest single cause for the most visible failures in the telecommunications industry in the last ten years or so. Company after company raised significant amounts of investment capital, and spent countless billions of dollars building expensive networks to carry international long distance traffic, that was dropping 5-10% *per month*. By the time some of these networks were capable of carrying traffic, the amount of volume needed to meet the business plan had often outstripped the capacity of the network by an order of magnitude, or more.

Lesson here? If you want to make money in telecom, make sure that you include some worst case pricing actions in your financial projections. Today's market conditions may not be the same as tomorrow's, so analyze the trends carefully.

Clearinghouses

Clearinghouses and Least Cost Routing carriers are relatively new phenomena, both primarily reliant on buying minutes from other carriers, and reselling those minutes. This is a business with very slim margins, and it is becoming increasingly unattractive because the dropping unit prices require proportionately greater volume just to maintain revenues. Most clearinghouses try to combine their own direct routes with the routes of others in order to average up the overall margins. While direct routes do offer some meaningful margins, they are also subject to swelling tides of competition, placing continual downward pressure on prices and margins.

There are many medium to small clearinghouses out there, who are eking out a living on the arbitrage business. But even the large, well-funded VoIP clearinghouses are on shaky ground. By and large, they have not been able to secure a firm market position. Mainly because most users can see only one difference between their services and those of the traditional carriers — a lower price. As traditional carriers lower prices, and the price differential lessens, these clearinghouses have more and more difficulty maintaining their customers.

Making money in an arbitrage oriented clearinghouse is a day by day event, a continual process of finding lower rates and new customers. If this proceeds to its logical conclusion, the new customer acquisition cost and churn could equal the narrow margins, making it impossible to achieve any sustainable level of profit.

Prepaid Calling Cards

This remarkably resilient industry has succeeded since its innovation in Italy in the 1980s. Until a few years ago, every telecom guru and pundit would predict that prepaid cards would meet their demise the following year. And then, every year the prepaid calling card market would increase, often dramatically. No serious prognosticator makes this prediction anymore, and the market proceeds to increase in both revenues and new applications.

There are three primary facets to the prepaid calling card industry, each of which has a different way of making money. We can call them the "3 Cs" of prepaid calling cards.

1. The Commodity Market

This is the largest segment, from a revenue point of view, and it's the most competitive. Go to a convenience store near a port, or in an inner city area, and you will see an array of calling card posters, each shouting its offer to you. (The experience is reminiscent of New Orleans, where street hawks hype their products.) The offers are very similar, and based on a promised number of minutes to given destinations.

"89 minutes to the Philippines for $10", "500 US long distance minutes for $5", "287 minutes to Brazil for $20" are the type of offer you see. If you do a little math, the offers are incredibly cheap, especially if you consider the inbound toll free costs, etc. In fact, if you simply divide the amount of money by the number of minutes, getting the cost per minute, it will look to be very cheap, suspiciously cheap. Hmmmm?

How do they do it? Take a look at the fine print at the bottom of the card, or on the poster, and you will see a series of terms and conditions that will offer a clue to what is going on. First, there will be surcharges, maintenance charges, payphone charges, minimum call durations and rounding up the minutes, and so on. All of these deduct from the balance available, and reduce the number of minutes. In theory, there should be at least one scenario that will allow the caller to get the advertised number of minutes. That is when the caller "burns" the card, and uses the card to make one call.

This is quite a gamble. If you make the first call, and Uncle Joe is not at home, but Cousin Abby answers the call, you lose. In fact, your 500 minutes might only be 50 minutes, or even less, after you make just one call. In fact, with maintenance charges, you might not have to make a call before money is deducted. Some cards even deduct the first maintenance charge before the first call, obviously reducing the number of available minutes for that call.

And, all of this is before any more nefarious schemes are put into place. For example, the infamous "50 second minute", where calls are charged a full minute for every 50 seconds. Some prepaid cards have also been known to deliberately disconnect calls if they go on too long, so the card holder will not have the opportunity to make even that one call. Bear in mind that I am not advocating any of these unethical practices, just advising you that they happen.

Prepaid calling card companies that manage their cards carefully can make considerable profits, even if they fulfill their promises. A knowledge of the markets and usage patterns will allow them to design programs that will be attractive to consumers while still affording a decent profit margin. The fact that there are few, if any, successful telecom companies left that have commodity prepaid calling cards as their primary product line, should be a warning to investors.

To consumer buying these commodities, *caveat emptor*!

2. The Convenience Market

This is probably the best market for prepaid calling cards, and the only one that still has major companies competing for business. This market is best exemplified by the vending machines at the major airports of the world. These convenience cards are sold, at premium rates, to consumers who are in immediate need of a calling card, like a foreign traveler. The rates are very high when compared to the commodity cards, but lower than a collect or operator assisted call from a pay phone.

These cards are also sold at convenience stores, hotels, airport shops, military bases, schools and buying clubs. Again, the rates are not eye-popping, but the terms are usually less onerous and not so one-sided in favor of the card provider. One rate cellular plans, offering "free" long distance calling, have eaten heavily into this market. Look at airports in countries with heavy cellular usage, like the US and the UK. In the US, there are estimates that in recent years 12 million pay phones have been, or are being, disconnected.

These cards have great margins, but usually a heavy portion of the profits have to be shared with the location where the cards are sold. The consumer gets a decent value, with few, if any, hidden costs and saves money over other calling methods. Investors will find reasonable margins, but the prepaid products may be hidden deep in the financial statements, since they are not primary products.

3. The Collectible Market

The final market for prepaid calling cards is the collectible market, where the primary reason for buying the card is not for the service that the card provides, but for the card itself. In fact, most collectibles put the card in a cellophane envelope and have a "scratch-off" covering the PIN number. The seller will then admonish the buyer that the card will diminish in value if taken out of the envelope. This means that if you actually use the card, you will reduce the value of the card, according to the seller. Of course, if the seller can convince the buyer not to use the card, their profits will be enhanced accordingly.

This has been an inconsistent market, and in the past even had magazines and trade shows dedicated to it. Although it has declined in popularity, there are still many fans, and cards still trade. If you decide to invest in cards, be forewarned, it is like any other collectible, and the market for cards can be very volatile. Calling card collection is not as well established as coin or stamp collecting, making it as risky as gambling, or playing the telecom stock market.

This overview of how money is made in the telecom business should give you an appreciation for the complexities involved in actually realizing a profit. Telecommunications is a very dynamic industry, one that has gone through deep and dramatic changes in the past few years. The prospects are that an even greater change in the future will occur. As with any endeavor, knowledge is power, so study any telecom opportunities carefully, looking hard for evidence of sustainable business prospects.

Chapter 36

A View of the Future

Next gen telephony is so new and changing so rapidly that it is difficult to accurately forecast precisely what is going to happen next, or what is going to happen at all. Witness our Bell Labs picturephone example earlier. Nearly every writer and analyst from 1960 to 1990, at one time or another predicted that by now every home would have a picturephone as its central communications fixture. Today, the hardware for this technology is there, and it's cheap. While there might be a slight problem with the bandwidth for full motion video in most homes, nearly every business has adequate bandwidth, and limited motion video can be sent over dial up lines in homes.

So why don't we all have picturephones as predicted?

There are several possible answers to the question of why picturephone never lived up to its promise, but for the purposes of prognosticating the future, the most important lesson is that the consuming public will (a) have the final say, and (b) will make its own decisions, regardless of how attractive the product or service might be to the vendors, investors and pundits. You might note that we also do not have flying cars, robots, or "intelligent homes", all of which are clearly within the capabilities of the current state of our technology. All of these have been commonly predicted by experts as well. There are some cost issues with these devices, but that could probably be mitigated by mass production.

But, once again, the consumer demand is just not there. You can overcome a lot of things when you try to market a new product, but at any time, the consumer in a free market will have the final say, and will often veto even the most compelling ideas.

So, with all the proper fanfare, and disclaimers, I present my top predictions for the future of telecommunications.

Prediction #1 — A PC is just a PC

You will not be making telephone calls over your computer anytime in the near future. Why? There is simply no need to use a thousand dollars worth of fragile, temperamental technology to do what can be done very conveniently on a device you can now buy for $10. And the $10 device is much easier to use, "grandma friendly", can be moved to another room easily, and doesn't crash. Why try to make calls over the Internet, calls that are typically low in quality, quirky and subject to interruption, noise and delay.

Nobody wants that. What people want is lower unit prices for long distance calls, and with the drastically lower prices that accompany open unfettered competition, who needs the Internet?

There is also a bandwidth issue in providing Internet telephony services to most homes, but DSL, cable and wireless are making great inroads here. Will the day ever come when all the communications services are delivered to homes and businesses on a strand of light? Maybe, but there are several issues that may preclude this. First, what is the advantage to the consumer for having a single provider of Internet, telephone, long distance, cable TV, and who-knows-what-else? As a consumer, I don't see it.

There is no economy of scale that I am going to enjoy by buying everything from one source. Wasn't this what the Bell System was broken up for? Isn't this the issue behind the global liberalizations we see? The reality is that, once a telecommunications provider has the leverage over the consumer, they will raise prices to make money. Sorry, but that is the truth. Look at our example of the price of an international long distance call to the Philippines in the previous chapter. Which service had the highest price, and highest margin, for the call? Cellular. Which service offers the least options on making international long distance calls? Cellular.

Coincidence? I don't think so. If you analyze why cellular providers (who usually give you no choices on the selection of domestic long distance carriers) have the highest price on international long distance calls, the answer is, "because they *can*." It is so obvious that, without competition, vendors will price to their advantage, that it's surprising how often we forget to remember this. Do you honestly believe that cable Internet access would be as "reasonable" as it is, if it did not have to compete against DSL and wireless?

OK, so the economy argument for Internet telephony does not stand, so what are the other reasons you would want to use your computer to make telephone calls? Convenience? Nope. Ease of use? Get serious! Reliability? Not with Windows. Advanced features? Hmmm, maybe. Granted, IP devices can provide some sexy, and maybe desirable new features, but we don't have to try to carry the actual voice signal over the public Internet. There is plenty of bandwidth available, and we can offer these features using SIP phones directly connected to other types of networks, and without the fear of congestion and delay.

Prediction #2 — The Bandwidth Glut Will Vanish, and Then Some.

Let's address the bandwidth issue for a moment, because one of the issues that led to the telecom crash of 2000-2001 was the incredible build out of fiber bandwidth around the world. Some pundits have predicted that there will be such a glut of bandwidth in the future, that the telecom sector will suffer for years.

Let me give you the benefit of Retske's Second Law of the Universe,

The applications used over any communications network will expand to fill all of the available bandwidth, and then some.

(If you are interested in Retske's First Law of the Universe, contact me. My beloved editor would not allow me to print it here.)

Look at the facts. I visited bulletin boards in the early 80s using a 1200 baud dial up modem, which was sufficient for the application. Then I got CompuServe, and it was slow, but a 9600 baud modem was able to stay up with it. Then the Internet became available, and 14.4 was enough, until the web pages starting using graphics, and 28.8 was enough, until 56K became widely available, so web pages added sound, video and moving graphics to fill the bandwidth. Last night I listened to an NFL football game over the Internet, and the quality was acceptable. The applications have expanded to fill all of the available bandwidth.

Look at cable TV, and the expansion of channels since the advent of TV, when two channels was a luxury. Today, there are hundreds of channels on cable and satellite TV. (I have to admit that UHF TV in the US seems to violate this principle, but the cost of building a UHF TV station was excessive and doomed it to fail to grow before it had a chance.)

Will the rapid expansion of bandwidth suppress revenue growth for the foreseeable future, as some analysts have suggested? Maybe, but not for long. Telecommunications commodity exchanges are already moving in to broker the bandwidth. Prices that initially dropped quickly are beginning to level off. In fact, in the second half of 2001, many international long distance rates in the US, the UK and much of Europe actually went up fractionally. What is happening here is that the prices are starting to sink to levels that are more in line with the actual costs. No one is going to sell anything for less than it costs them. Not for very long anyway, but for years the prices for many telecom services were held artificially high by the monopolistic incumbent carriers. Now that competitive pressures are coming to bear, the prices should drop to a level that is more normal for a competitive model.

Prediction #3 - Wireless Will Rule in Growth

OK, pop quiz! What is the only sector of telecom that still does not have significant competition, anywhere in the world? Wanna hint? What is the most profitable sector? No, not cellular, it has competition, nearly everywhere. Last clue — without it, most of today's telecom services could not exist. Give up? How about local service? Remember? Last mile. LEC. CLEC (giggle, giggle). BOC. Baby Bell. License to steal.

There have been some valiant efforts at offering competitive last mile services, but they have fallen short of the mark, so far. Note that we are differentiating Last Mile Services (LMS, an acronym I just made up!) and CLEC, BLEC, etc. services. The difference is that an LMS has to own the actual facilities used to deliver service to the user. Many CLECs resell the Incumbent LEC (ILEC) services, and do not actually construct anything for the last mile.

Just in case you are not familiar with the term, "last mile" is a reference to the connection from the central office or switching point to the users' premises. The term may be a misnomer, since the actual distance involved is probably not exactly a mile. In most cases, it means that the facilities must be strung over, usually, public rights of way. Electric companies, cable television companies and other utilities have made stabs at entering the LMS market, but so far without a great deal of success. Business users in dense, inner city locations, may find fiber providers that offer LMS in competition with the LEC, but this is primarily directed at the upper echelon of users.

The last great hope is wireless. Wireless offers ubiquitous availability. Look at cellular mobile telephone services, which are available to over 90% of the world's population. But the cost of cellular calls, coupled with the miserable bandwidth (28.8 kbps is a stretch), will relegate traditional cellular to its current status.

However, 3rd Generation, or 3G, wireless technology has the potential for offering a service that could compete with LEC LMS service. Perhaps not in its current form, but the technology is evolving rapidly, and could find itself in a battle with LECs. Because 3G does not require any physical connection — wires, cables or fibers — it avoids all the legal and franchise issues involved with crossing public corridors.

Since LEC services have become the new cash cow following the demise of monopoly long distance services, it will be the target for competitive efforts. Watch for wireless LMS services. They may be coming to the airways near you!

Prediction #4 — Entrenched Monopolies Will Continue the Fight

All right, this may be an easy one, but I deserve it!

The entrenched monopolies include the Regional Bell Operating Companies (RBOCs), PTTs, LECs and others with a regulated service that virtually guarantees a profitable business. For example, here in the US, the RBOCs and LECs. They were offered an incentive for losing local exchange market share. The 1996 Telecommunications Act included a provision that allowed them to offer lucrative long distance service, but only if they could demonstrate that they had effective local competition.

It is unimaginable that any business will willingly give up a lucrative, profitable business. What has happened here is a battle of lawyers and words that will logically culminate in all the RBOCs getting authorization to offer long distance services. It cannot end any other way. Just as the battle for international long distance competition globally will ultimately end up with formally approved competition worldwide. But if you think the battle ends with the authorization, you are not looking at the history.

In the US, when MCI first got authorization to offer long distance, (actually private line competition), the war had just began, and it is still being fought today. Market dominance, even in the absence of a formally granted monopoly, is a powerful weapon. Look at personal computer operating system software. Nobody gave Microsoft a monopoly. They earned it. Regardless of how they got it, you think they will NOT use every device and resource at their disposal to hold on to what they have?

Our friends in the "former" monopolies are no different. Survival is one of the strongest human instincts, so they are undoubtedly pleased that some of their potential strategic competitors have become as rare as neckties on a golf course. You can bet that they will also use every means at their disposal to frustrate any other competition that might happen to come their way.

Prediction #5 — Telephone Bills Will Keep Going Up

There is a myth going around that telephone bills are going down. Wait a minute! Didn't we say earlier, on several occasions that prices were dropping? Yes, we did, and it is true that *unit* prices are dropping for the vast majority of products and services. But, the world has become more reliant on information services, and use more of them everyday. Lower unit prices make the services accessible to more people, and make the use of the services more economical to everyone.

Don't believe it? Take a look at the US long distance market. Every year since the mid 70s, AT&T has lowered the unit price, in response to competitive pricing pressures. And, every year, AT&Ts revenues in long distance went up.

With new services being introduced everyday, this counter-intuitive prediction is a lead pipe cinch!

Prediction #6 — Mergers and Acquisitions Will Accelerate

Just when you think you have seen it all, it is going to get worse, or better, depending on your perspective. There are literally hundreds of major telecom companies, worldwide. PTTs, local service companies, mobile companies, etc. All of these are under extreme pressure to increase sales every year. Flat revenues, holding your own, doesn't meet investor and stockholders' expectations, so, Mr. or Ms. CEO, if you can't do it for us, your replacement will!

The fastest way to increase revenues is the "roll 'em up" method. Acquire someone, preferably a competitor, killing two birds with one stone, and your life as a CEO will be better. (Or at least, longer.) There are plenty of targets out there, among the startups, the national companies, and so on. If you are a $100 million company, and acquire a $25 million company, you just increased revenues by 25%, a major accomplishment.

Look for this trend to continue.

Prediction #7 — Picturephone Will Fail Again.

Believe it or not, someone will likely see this as a significant new opportunity, and will try to capitalize on this "exciting new market." As recently as 1997, AT&T actually tried, again, so this is not such a frivolous prediction. By the way, this attempt failed, despite some significant promotion and advertising.

One good thing: if anyone does try again, they will not have to create much new in the way of adverting copy and marketing materials. There is plenty to draw on from the past.

My prediction as to how this new effort will turn out? Guess!

Prediction #8 — Telecom Investments Will Finally Begin to Pay Off

There is nothing to indicate to anyone that the telecommunications industry is likely to fade or go away, right? In fact, exactly the opposite is true. The probability is that the world will become ever more dependent on the flow of information. Compelling new devices and services will lure consumers into inching their telecommunications expenditures up a bit more. The Internet will take on more importance, once it settles down a bit.

All in all, telecommunications has a bright future. Having said that, I will also say that some, maybe many, of the companies that are at the forefront now, will not make it through the next few years. At least, not as they now exist. They may get acquired, merge, or just die, but 10 years from now, they may not be around.

But, if you pick the right investments, you may be able to make a lot of money in telecommunications. You have to have a great deal of knowledge, experience and luck, but it can be done. There is money to be made for those with the persistence to last it out.

I, for one, hope you do!

Appendix I: TCP/IP

One of the key enabling technologies behind the Internet is TCP/IP, a protocol that allows networked computers to communicate with each other whether they are part of the same network or are attached to separate networks. One computer could be a Cray and the other a Macintosh, and TCP/IP will facilitate communications between them. TCP/IP is a platform-independent standard that bridges the gap between dissimilar computers, operating systems, and networks.

TCP/IP is actually a collection of several different protocols that TCP/IP hosts use to exchange information. Let's take a look at some of those protocols and see how they contribute to the over all TCP/IP structure.

TCP/IP Basics

Computers run on operating systems, like UNIX or Windows, that allow applications, like word processors to function and provide other services to the user. The operating system operates the hardware of the computer, and controls any communication between the computer and any other computer. To allow the exchange of data between computers, networks are created, often with proprietary protocols like SNA, SPX/IPX, or NetBUEI.

Computers on a peer-to-peer network are connected to each other through a hub, a device that allows electrical connections to be made between the Network

Interface Card (NIC) installed in each computer. The most common NIC is one that supports Ethernet connections, either twisted pair, 10base-T, or coaxial cable. These cards will allow a minimum transmission speed of 10 megabits per second (mbps), or 100 mbps in an increasingly common arrangement.

These protocols are not able, often by design, to communicate with dissimilar networks. TCP/IP, because of its universal availability across nearly every computer platform and operating system, is the "glue" that allows diverse networks to be connected to each other in private intranets, or over the public Internet.

TCP/IP stands for Transmission Control Protocol/Internet Protocol. In simple terms, a protocol is a mutually agreed standard enabling computers or networks to exchange data with each other. TCP/IP is actually a combination of several protocols, or a protocol suite, two of which are TCP and IP, the two principal protocols that facilitate communications over networks.

TCP/IP software installed on a computer or a network, provides implementations of TCP, IP, and other members of the TCP/IP family that are specific to the hardware and operating system of the platform on which they are loaded. Typically, this collection of software also includes some high-level applications, like FTP (File Transfer Protocol), which permits users to perform network file transfers between computers on different networks.

The networks that comprise an intranet are physically connected to each other through a gateway device called a router, or an IP router, that provides for the interconnection of different hubs. A router transfers packets of data from one network to another, interpreting or translating addresses as needed, and picking only those packets that are destined for other networks, or subnetworks. Information travels on a TCP/IP network in discrete units called IP packets or IP datagrams. TCP/IP software makes each computer attached to the network a sibling, or a pier, to all the others. In essence, it hides the routers and underlying network architectures.

TCP/IP grew out of research funded by the U.S. government's Advanced Research Projects Agency (ARPA) in the 1970s. It was developed so that research networks around the world could be joined together to form a virtual network, known as an internetwork. The original Internet was formed by interconnecting an existing conglomeration of networks, known as ARPAnet, which used TCP/IP to communicate. This "internet" would eventually become the backbone of today's Internet.

This adds a layer of transparency, and makes everything seem like one big network. Just as connections to an Ethernet network are identified by 48-bit Ethernet IDs, connections to an intranet are identified by 32-bit IP addresses, which we express as dotted decimal numbers for convenience (for example, 128.10.2.3). Given a remote computer's IP address, a computer on an intranet or the Internet can send data to the remote computer as if the two were part of the same physical network.

TCP/IP also provides a solution to the problem of how two computers attached to different physical networks can be part of the same intranet, and can exchange data with each other. The solution comes in several parts, with each member of the TCP/IP protocol suite filling in one piece of the puzzle. The most fundamental TCP/IP protocol, IP, allows for an important function called routing, which provides for the transmission of IP datagrams across an intranet, choosing a path that datagrams will follow to get from A to B. Routers allow these datagrams to "hop" between networks.

TCP divides data streams into chunks called TCP segments and transmits them using IP. In most cases, each TCP segment is sent in a single IP datagram. If necessary, however, TCP will split segments into multiple IP datagrams that are compatible with the physical data frames that carry bits and bytes between hosts on a network. Because IP doesn't guarantee that datagrams will be received in the same order in which they were sent, the TCP protocol provides for the receiving end to reassemble the datagrams to form an uninterrupted data stream. FTP and telnet are two examples of popular TCP/IP applications that rely on TCP. Building on this transmission routing scheme, TCP has an even higher-level protocol that allows applications running on different hosts to exchange data streams.

Other TCP/IP protocols play supporting roles in the operation of TCP/IP networks. For example, the Address Resolution Protocol (ARP) translates IP addresses into physical network addresses, such as Ethernet IDs. A related protocol, the Reverse Address Resolution Protocol (RARP), does the opposite, converting physical network addresses into IP addresses. The Internet Control Message Protocol (ICMP) is a support protocol that uses IP to communicate control and error information regarding IP packet transmissions. If a router is unable to forward an IP datagram, for example, it uses ICMP to inform the sender that there's a problem.

Another important member of the TCP/IP suite is the User Datagram

Protocol (UDP), which is similar to, but more primitive than TCP. TCP is a "reliable" protocol because it performs the error-checking and handshaking necessary to verify that data makes it to its destination unchanged. UDP is an "unreliable" protocol because it doesn't guarantee that datagrams will arrive in the order in which they were sent or even that they will arrive at all. If reliability is desired, it's up to the application to provide it. Still, UDP has its place in the TCP/IP universe because of lower overhead, and a number of applications use it. The SNMP (Simple Network Management Protocol) application provided with most implementations of TCP/IP is one example of a UDP application.

TCP/IP Architecture

Network designers often use the seven-layer ISO/OSI (International Standards Organization/Open Systems Interconnect) model when discussing network architectures. Each layer in the model corresponds to one level of network functionality. At the bottom sits the physical layer, which represents the physical medium through which data travels—in other words, the network cabling. Above that is the data-link layer, whose services are provided by network interface cards. The uppermost layer is the application layer, where application programs that use network services run.

As a unit of data flows downward from a network application to the network interface card, it travels through a succession of TCP/IP modules. At each step along the way, it is packaged with information required by the equivalent TCP/IP module on the other end. By the time the data makes it to the network card, it has been transformed into a standard Ethernet frame, assuming the network is an Ethernet network. The TCP/IP software on the receiving end recreates the original data for the receiving application by grabbing the Ethernet frame and passing it upward through the TCP/IP stack. (One of the best ways to understand the guts of TCP/IP is to use a "sniffer" program to look inside frames flying around the network for the information added by the various TCP/IP modules.)

To picture what role TCP/IP plays in real-world networks, consider what happens when a web browser uses HTTP (HyperText Transfer Protocol) to retrieve a page of HTML data from a web server attached to the Internet. The browser uses a high-level software abstraction called a socket to form a virtual connection to a server. To retrieve a web page, the browser sends an HTTP GET command to the server by writing the command to the socket. The socket software in turn uses TCP to send the bits and bytes comprising the GET command to the web server. TCP segments the data and passes the individual segments to the IP module, which transmits the segments in datagrams to the web server.

If the browser and the server are running on computers connected to different physical networks (as is usually the case), the datagrams go from network to network until they reach the one to which the server is physically connected. Ultimately, the datagrams are delivered to their destination and reassembled so that the web server, which receives chunks of data by performing reads on its socket, sees a continuous stream of data. To the browser and the server, data written to the socket at one end shows up at the other end, as if by magic. But underneath, all sorts of complex interactions have taken place to create an illusion of seamless data transfer across networks, thus giving rise to the term "virtual connection."

And that's what TCP/IP is all about: turning lots of small networks into one big one and providing the services that applications need to communicate with each other over the resulting internet.

In Review

There's much more that could be said about TCP/IP, but here are three key points:

1. TCP/IP is a set of protocols that permit physical networks to be joined together to form an internet. TCP/IP combines the individual networks to form a virtual network in which individual hosts are identified not by physical network addresses but by IP addresses.

2. TCP/IP uses a multilayered architecture that clearly defines each protocol's responsibilities. TCP and UDP provide high-level data transmission services to network application programs, and both rely on IP to transmit packets of data.

3. IP is responsible for routing the packets to their destination. Data moving between two applications running on Internet hosts travels up and down the hosts' TCP/IP stacks. Information added by the TCP/IP modules on the sending end is stripped off by the corresponding TCP/IP modules on the receiving end and used to re-create the original data.

Appendix II: Bandwidth

Facility	Bandwidth	Voice Channels	Description
DS0	64 Kbps	1	Basic Voice Channel
T1, DS-1	1.544 Mbps	24	North America,
E1, DS-1	2.048 Mbps	30	Europe, Asia, South America
T2, DS-2	6.312 Mbps	96	4 T1s
E2	8.448 Mbps	120	4 E1s
Ethernet	10 Mbps		Basic LAN
E3	34.368 Mbps	480	16 E1s
T3 or DS3	44.736 Mbps	672	28 T1s
OC-1, STS1	51.840 Mbps		Basic Sonet (STS is Electrical equiv.)
Fast Ethernet	100 Mbps		Fast LAN
OC-3, STS3	155.520 Mbps		3 OC1s
OC-12, STS12	622.08 Mbps		12 OC1s
OC-48	2.488 Gbps		48 OC1s
OC-96	4.976 Gbps		96 OC1s
OC-192	10 Gbps		192 OC1s
OC-255	13.21 Gbps		255 OC1s

Appendix III:

Recommended Reading

Note: These are books that I think are very relevant to subjects in this book, and are among my favorites. There are two reasons that this is not a long list. (1) Long bibliographies may look impressive, but they make it difficult for the reader to find an applicable book, (2) I don't read a lot of books.

The International CallBack Book, An Insider's View, Gene Retske, 340 pages (October 1995) CMP Books; ISBN: 0936648651 - My personal favorite. Out of print, but contact me if you can't find one.

Newton's Telecom Dictionary, Harry Newton/Ray Horak, 787 pages 17th edition (March 2001) CMP Books; ISBN: 1578200695 - If you don't already have at least one of these, get out of the business.

Communications Systems and Networks, Ray Horak, 697 pages 2nd edition (January 15, 2000) Hungry Minds, Inc; ISBN: 0764575228 - Most complete technical tutorial/reference in understandable language.

Carrier Grade Voice Over IP, Daniel Collins, - 496 pages (September 22, 2000) McGraw-Hill Professional Publishing; ISBN: 0071363262 - To understand the details of VoIP and legacy network interconnections, this is the book.

Voice Over IP, Mark A. Miller, 296 pages Book & Cd-Rom edition (February 15, 2000) Hungry Minds, Inc; ISBN: 0764546171 -Basics of VoIP for the masses, includes all relevant RFCs on disk.

Frames, Packets and Cells in Broadband Networking, William A. Flanagan, 216 pages (June 1991) CMP Books; ISBN: 0936648317 - Understanding what the packets contain, and how they communicate, is vital to understanding the technology.

Telecosm, George Gilder. 351 pages (September 11, 2000) Free Press; ISBN: 0684809303 - Very informative reading, even if I don't agree with a lot of his analysis and predictions. We'll see, George!

Louisiana - Real and Rustic, Emeril Lagasse, 347 pages (September 1996) William Morrow & Co; ISBN: 0688127215 - Totally irrelevant to telecom, but delicious!

Irwin Handbook of Telecommunications, James Harry Green, 845 pages 4th edition (February 17, 2000) McGraw-Hill Professional Publishing; ISBN: 0071355545 - An excellent tutorial of all technical aspects of telecommunications.

Glossary

1 Wiltshire This is one of the biggest and oldest "telecom hotels" in the US. It is located at 1 Wiltshire Boulevard in Los Angeles, but is called "1 Wiltshire" for short. It is probably the largest "meet me" point on the west coast of the US, and is the gateway to the Pacific Rim.

1996 Telecommunications Act This US act was passed in 1996, as an extension of the original Telecommunications Act of 1934, passed (ironically) in 1934. A lot happened between 1934 and 1996, so this Act was intended to formally establish competitive structures and to foster a free market environment in the US.

60 Hudson Street This is the other big meeting point and collocation facility in the US, and is probably the largest such facility in the world. Located in the Wall Street area of Lower Manhattan, it has become the model for collocation and meet me peering around the world. It is also a haven for union activity, under the table deals, graft and fraud. BEWARE!

Accounting Rate The amount agreed upon by two national carriers as the baseline rate per minute for charging each other for the traffic they exchange. Historically, this number was not based on anything but what the two correspondents thought they could get away with.

ACH Automated Clearing House. A form of Electronic Funds Transfer where an amount of money is transferred from one bank account to a different bank account, usually to pay a bill.

Aggregator One of the companies formed in the late 80s that used carrier tariffs to combine the bills of unrelated customers to achieve discounts, which they partially shared with the customer. Also known as an "aggrevator".

ANI Automatic Number Identification. A service where the called party gets the telephone number of the party that is calling them. The most visible example of ANI today is Caller ID.

Arbitrage The business of buying and selling traffic without adding value. It is a commodity trade, making profit by buying time at a price that is often only marginally lower than what was paid for it.

ARP Address Resolution Protocol. The low level service of TCP/IP that maps IP addresses to the Ethernet NIC physical address.

ASP Application Service Provider. Ostensibly, a company that offers an application over the Internet. These applications range from simple telecom services to high end payroll and accounting. The role of an ASP is largely undefined.

Bell System The former AT&T conglomerate before it was broken up in 1983. Comprised 7 regional companies, AT&T long distance (Long Lines), Western Electric, Bell Labs and others. Also had minority interested in Southern New England Telephone and Cincinnati Bell.

Callback Also known as International Callback, call re-origination and others. The use of existing network facilities in an application (ASP?) that allows a service provider to offer US dialtone and prices to overseas customers. Part of the service involved an unanswered trigger call, generating a great deal of controversy.

Carrier Or Common Carrier. A facilities based telecommunications service provider, which offers its services to the general public.

Carterphone Decision A legal decision by the US Federal Communications Commission in 1968 that overturned an AT&T attempt to control the use of its equipment. The implications of this decision led to the liberalization of the US telecommunications market, which is spreading throughout the world.

CCITT Comité Consultatif International Téléphonique et Télégraphique, or International Consultating Committee on Telephony and Telegraphy. The previous name of the International Telecommunications Union, ITU. *See ITU*.

Circuit Switched The routing and connection of telephone calls by actually switching the termination of the call by various means. A circuit constructed in this manner is dedicated from end to end for the duration of the call.

CLEC Competitive Local Exchange Carrier. A telecommunications service provider that offers local, i.e. dialtone, service, in competition with the established Local Exchange Carrier. Ultimately, all LECs will become CLECs (or maybe all CLECs will become LECs?).

CO Central Office. The local switching and cable termination point for LECs.

Collection Rate This is the charge that a national telecommunications carriers charges its end users for international calls. It has no relationship to the Accounting Rate, but is obviously influenced by it.

CTI Computer Telephony Integration. The first step in the converging of voice and data communications.

Datagram A single packet of information sent over a network. Usually part of a complete data stream that has been split into pieces to expedite its transmission. It is the basic unit of information in TCP/IP.

DID Direct Inward Dialing. A group or range of telephone numbers that share the same group of trunks to actually complete calls.

DNS Domain Name Server, or Domain Naming Service. The service provided on the Internet to translate alphabetic names used by browsers to the actual 32 bit binary addresses used on the Internet. A DNS server is a server on the Internet which provides DNS service.

DS0 Digital Signal, level 0. Equal to 64 kilobits per second, the smallest basic voice grade channel (using PCM).

DS1 Digital Signal, level 1. A T-1 circuit, comprised of 24 DS0s or 1.544 MBps.

E-1 The "other" T-1 standard, used in most of the world outside of North America, Japan and Hong Kong. It has a bandwidth of 2.048 Mbps, and is divided into 32 channels. In voice service, 2 channels are used for signaling, leaving 30 available for voice channel Ds0s.

ENUM The word "ENUM" refers to the IETF protocol that takes a complete, international telephone number and resolves it to a series of URLs using a Domain Name System (DNS)-based architecture.

ESS Electronic Switching System. A telephone switch that is totally electronic, with stored program control. The generation that replaced electro-mechanical switching systems, like #5 Crossbar and Step-By-Step.

Ethernet A Local Area Network system that uses twisted pair or coaxial cable to interconnect computers in a network.

European Union The fifteen countries that have banded together to form a loosely unified Europe. Currently the countries are: Austria, Belgium, Denmark, Finland, France, Germany, Greece, Ireland, Italy, Luxembourg, The Netherlands, Portugal, Spain, Sweden and the United Kingdom. The significance to telecommunications are the mandate on market liberalization, and the common currency, called the Euro.

Exchange Rate The dynamic, ever changing ratio at which the currencies of different countries are exchanged.

Facility A term that refers to the various elements of telecommunications services - network connections, switches, POPs, etc. It is a rather all-encompassing term, and not clearly defined.

FCC Federal Communications Commission. The United States federal telecommunications regulatory authority.

GATT General Agreement on Tariffs and Trade. The initiative that preceded the World Trade Organization.

Geosynchronous Satellites that are placed in orbit 22,300 miles above the Equator, and are made to circle at the same rate as the Earth's rotation, giving the appearance of being stationary.

Google One of the best new search engines. They don't hold auctions, sell ISP connections, offer free email or span the tar out of users. Highly recommended.

GSM Groupe Speciale Mobile, now known as Global System for Mobile Communications (GSMC). The cellular mobile system in use in most of the world, except North America.

H.323 The ITU standard that describes how VoIP devices can communicate.

HTML HyperText Markup Language. Authoring software language used to produce World Wide Web pages.

HTTP HyperText Transfer Protocol. This is the protocol that is used by web servers to communicate with web browsers.

IETF Internet Engineering Task Force. The primary planning and engineering body that develops and maintains TCP/IP standards for the Internet.

Internet When used with a capital "I", it refers to the public Internet, the global and pervasive computer network, open to the general public. When used with a lower case "i", it refers to any interconnection of two or more distinct computer networks, and is private in nature.

Internet Telephony The technology of conducting a telephone call over the Internet.

ISO International Standards Organization. Also known as IOS, International Organization for Standardization. The Geneva based UN organization that sets technical standards. Best known in telecommunications for its 7 layer model, called OSI

ISP Internet Service Provider. A telecommunications service provider offering one or more access methods for users to connect with the Internet.

ISR International Simple Resale. The use of international private lines to create the transmission paths for international voice traffic.

ITU International Telecommunications Union. A permanent UN agency charged with setting technical standards and overseeing telecommunications. The role and influence of the ITU have diminished since the advent of competition, and they now mainly produce telecom mega-trade shows and publish information.

IVR Interactive Voice Response. The technology that allows telephony users to communicate with computers. Uses touchtones or voice recognition to receive and interpret user input, and speaks recorded or synthesized speech back. Voice Mail, Automated Attendant, and other applications are examples of IVR.

LAN Local Area Network. A computer network that is local in scope, and monolithic in protocol.

Least Cost Routing Carriers or users who have two or more alternate routes may select the least expensive route for a given call.

LEC Local Exchange Carrier. The telephony service provider that offers local connections, the "last mile", dialtone and local telephone calls. *See CLEC.*

LEO Low Earth Orbit. A satellite that is relatively low (400-600 miles above the Earth) in orbit. It loses the advantage of being geosynchronous, but requires much less power, and does not have the disadvantage of excessive propagation delay.

Lucent Technologies The world's biggest telecommunications manufacturer. Created by the spin off from AT&T of Bell Labs, Western Electric and a field sales and service structure.

MC Multipoint Controller. An endpoint that manages conferences between 3 or more terminals or gateways.

MCI Microwave Communications Inc. The first strategic competitor to AT&T. Through its activities, especially those of a legal nature, its founder, the late William McGowan, was able to force the first significant market opening in telecommunications.

MCU Multipoint Control Unit. An MC and an MP combined together to form a basic switching unit for VoIP.

MGCP Multimedia Gateway Control Protocol.

Microsoft Obviously the creator of the predominant PC operating system, Windows. If you thought that AT&T was an aggressive monopolist, just wait until you see what Microsoft has in mind!

MP Multipoint Processor. Operates under the control of the MC and controls the actual data streams.

NANP North American Numbering Plan. The planning and coordination of North American area codes, now administered by Lockheed Martin, formerly a part of AT&Ts Bellcore.

Natural Monopoly An excuse that AT&T created to justify its death grip on telecommunications prior to Carterphone and the resultant opening of the telecommunications market. Since AT&T had never been granted a legal monopoly, it claimed that telecommunications should "naturally" be a monopoly, and protected from competition.

Next Gen Telephony A loose description of all the non-legacy telecommunications techniques, including VoIP, voice over DSL, and all forms of packet switching.

NIC Network Interface Card. The piece of computer hardware that interfaces between a computers internal bus and a computer network.

OCC Other Common Carrier. An old term, originally created to differentiate between the activities of a common carrier, like AT&T and a specialized carrier, supposedly offering non-traditional telecommunications services.

OSI Open Systems Interconnection. *See ISO.*

Packet Switched

The new technique of routing and connecting telephone calls by sending packets of digitized voice in chunks (datagrams) back and forth.

PBX Private Branch eXchange. A local telephone system that provides central office type services within a single location of an enterprise.

POP Point of Presence. The physical place where a carrier installs his switch and terminates any network connections. In the old days, this was a toll office or a central office, today, it could be as modest as a router in a rack at a co-location facility.

PSTN Public Switched Telephone Network. The worldwide telephone network, accessible by telephones all over the world.

PTT Post, Telephone and Telegraph. The name was originally given by the British to their telephone company, and became the name for the national telephone monopoly anywhere.

QoS Quality of Service. The level of technical quality provided by a service provider to their customer.

Refile The redirecting of traffic from one country to another through a third country, for the purpose of getting the lower rates offered.

Resale The business of acquiring facilities or services of primary carriers, and selling them to customers.

Settlement Rate The actual amount that one international carrier has to pay the other correspondent for the excess of traffic. The settlement rate is equal to one half the Accounting Rate. If the number of minutes is EXACTLY equal in both directions, neither party has to pay.

SIP Session Initiation Protocol. A telecommunications protocol originally designed to allow two IP devices to link up over a transmission circuit. It has been focused on supporting Internet Telephony.

SLA Service Level Agreement. The primary agreement between a telecommunications service provider and a user or reseller, which defines the service provided, and a way of measuring the quality of service provided.

Specialized Common Carrier An early form of carrier that was intended to offer telecommunications services not provided by incumbent carriers like AT&T. In reality, they often just provided competitive services.

Speed Dialing A method for dialing complete sequences of telephone numbers and control codes by dialing a abbreviated number of digits.

SS7 Acronym for Signaling System 7. SS7 is the signaling protocol that has become the worldwide standard. SS7 provides many services in addition to signaling, enabling enhanced features like Caller ID. SS7 is an out of band signaling service, allowing maximum efficiency of telecommunication channels.

Switch A telecommunications term for a PBX, central office, or any other device that facilitates ad hoc communications between parties. In the old days, a Central Office was comprised of physical switches, giving rise to the name "switch". Even though these devices are almost entirely solid state today, the term has stuck, and has been extended to the function provided by routers and gateways.

T-1 Technically, this service is actually called T1.5, meaning that it provides a 1.544 Megabit per second transmission speed. T-1 travels over a single pair of copper wires, and is the standard for carrier communications services in the US, Canada, Japan and Hong Kong. T-1s can be connected directly to most PBXs, routers and data communications equipment.

Tariff The public filing by a regulated telecommunications company, detailing its service offerings, conditions and pricing. These tariffs, filed with the Federal Communications Commission in the US, and the PTT or regulatory authority elsewhere, represent the formal offering by a common carrier to the public in general. Prior to the onset of competition, tariffs guaranteed that every customer would be treated equally.

TCP/IP Telecommunications Communications Protocol/Internet

Protocol. This is a networking protocol that provides communications between interconnected networks, even if the internal networking protocol or operating systems are different. TCP/IP is the major enabling technology of the Internet. See the appendix section on TCP/IP for a detailed discussion.

TDM Time Division Multiplexing. A method for carrying multiple communications paths over one physical connection by sequentially allocating the available bandwidth in short chunks of time.

Toll Office An old term for a long distance switch.

Trunk A voice grade communications path. Usually describes a physical pair of copper wires.

Unified Messaging A system for consolidating all the various message sources - voice, fax, email, paging, etc - into a common location where it can be accessed as one source. Easier said than done!

UNIX The multi-user computer operating system created by AT&T in the early 1970s. UNIX or its derivatives are the basis for most of the AT&T and Lucent switching products, and runs most of their network. Today, UNIX is the predominant operating system running Internet servers.

URL Universal Resource Locator. The human language address used on the Internet, that is used to direct the inquiry to the proper IP address and resource.

VoIP Voice Over IP (Internet Protocol). The routing of calls over internets (lower case "i") or THE Internet using data communications connected devices. This is a really horrible definition, and you should read the entire book to see what it is all about.

WAN Wide Area Network. A computer network using common carrier provided circuits to extend the LAN over greater distances.

WTO World Trade Organization. The United Nations body charged with the mission of facilitating trade among the nations of the world. The successor to GATT, which in 1997, reached an accord that is designed to open telecommunications markets around the world.

Milton Keynes UK
Ingram Content Group UK Ltd.
UKHW022107141024
449569UK00031B/1806